Mathematics in Mind

SpringerBriefs in Cognitive Mathematics

The monographs and occasional textbooks published in this series tap directly into the kinds of themes, research findings, and general professional activities of the **Fields Cognitive Science Network**, which brings together mathematicians, philosophers, and cognitive scientists to explore the question of the nature of mathematics and how it is learned from various interdisciplinary angles. Themes and concepts to be explored include connections between mathematical modeling and artificial intelligence research, the historical context of any topic involving the emergence of mathematical thinking, interrelationships between mathematical discovery and cultural processes, and the connection between math cognition and symbolism, annotation, and other semiotic processes. All works are peer-reviewed to meet the highest standards of scientific literature.

Stacy A. Costa, ON Institute for Studies in Education
University of Toronto
Toronto, ON, Canada

Marcel Danesi, Department of Anthropology
University of Toronto
Toronto, ON, Canada

SpringerBriefs present concise summaries of cutting-edge research and practical applications across a wide spectrum of fields. Featuring compact volumes of 50 to 125 pages, the series covers a range of content from professional to academic. Briefs are characterized by fast, global electronic dissemination, standard publishing contracts, standardized manuscript preparation and formatting guidelines, and expedited production schedules.

Typical topics might include:

- A timely report of state-of-the art techniques
- A bridge between new research results, as published in journal articles, and a contextual literature review
- A snapshot of a hot or emerging topic
- An in-depth case study
- A presentation of core concepts that students must understand in order to make independent contributions

SpringerBriefs in Cognitive Mathematics showcases research in the study of math cognition as well as the relation of math to other faculties, such as language, music, and art, among others. As a subseries of **Mathematics in Mind** , the research published in this series falls under the aegis of the Fields Cognitive Science Network, which brings together mathematicians and cognitive scientists to explore the nature of mathematics from various interdisciplinary angles. All works are peer-reviewed to the highest standards of scientific literature.

Yair Neuman

Mindmatics

A Nexus of Ideas

 Springer

Yair Neuman
Dept of Cognitive and Brain Sciences
Ben-Gurion University
Be'er Sheva, Israel

ISSN 2522-5405   ISSN 2522-5413   (electronic)
Mathematics in Mind
SpringerBriefs in Cognitive Mathematics
ISBN 978-3-031-74954-4   ISBN 978-3-031-74955-1   (eBook)
https://doi.org/10.1007/978-3-031-74955-1

This Springer imprint is published by the registered company Springer Nature Switzerland AG
The registered company address is: Gewerbestrasse 11, 6330 Cham, Switzerland

If disposing of this product, please recycle the paper.

The book is dedicated to my granddaughters Aviv and Shoham.

Preface

כרטיס הכניסה להיכל הידע – הקושיא" מר שושני"

"The admission ticket to the temple of knowledge is the Kushiya." Monsieur Chouchani[1]

Kushiya (Hebrew קושיא) is an ancient Talmudic Aramaic term that has no exact translation into English. ChatGPT-4 translates it as a question. However, it is not a question. A Kushiya does not express a lack of knowledge that needs to be satisfied by an answer, but knowledge that contradicts a certain given fact and, therefore, a discrepancy that needs to be resolved (Wikipedia). Monsieur Chouchani,[2] the mysterious Jewish teacher who wrote the above-cited statement, proposed it as the key to what he poetically described as "the temple of knowledge." A first and simple reading of this statement seems to be clear. Discrepancies are crucial for the advancement of knowledge. Therefore, instead of silencing criticism and avoiding discrepancies, we must embrace them wholeheartedly. For instance, imagine yourself as a mathematician struggling to solve an equation, then suddenly realizing there is a solution. However, the solution involves the square root of -1, a strange object that has no right to exist in your mathematical world. How can you resolve this discrepancy? In fact, resolving this discrepancy was an important step forward as it introduced the idea of *imaginary numbers* into mathematics. Identifying and resolving discrepancies is necessary for progress, an idea that fundamentalists always find difficult to accept.

Seeking discrepancies is an important process, no doubt. However, as I read Chouchani's statement, his insight should not be interpreted as merely pointing to the importance of identifying and resolving discrepancies, paradoxes, or contradictions to advance our knowledge. To clarify this point, let me turn to the meaning of Kushiya.

Its meaning is grounded in *Kashe* meaning *hard*. In what sense is an issue "hard" or "difficult"? If it is a discrepancy, it can be bridged. If it is a paradox, such as Russell's Paradox, then it can be resolved. If it points to incompleteness, as in the

---

[1] http://www.chouchani.com/

[2] https://en.wikipedia.org/wiki/Monsieur_Chouchani

case of Godel's Incompleteness Theorem, then … let it be. However, my understanding is that Chouchani is pointing to something different. Indeed, he seems to emphasize the importance of identifying discrepancies through a never-ending process of interpretation. However, saying that identifying and resolving discrepancies may advance our knowledge is a rather trivial statement, and I doubt whether a genius like Chouchani would just be emphasizing the trivial. I consider his statement a Kushiya and provide my interpretation.

Careful attention should be given to the words used in the statement. We start with the Hebrew word ידע, which translates to "knowledge." In Genesis, we are first introduced to the word דעת (sometimes translated to English as "mind") in the context of the *Tree of Knowledge* (the tree of the "Daat"). The Tree of Knowledge is an inexact translation from Hebrew. The tree is originally described as the Tree of *Knowing* Good and Bad. One of the greatest Jewish thinkers, Maimonides (1138–1204), interprets the Biblical story by suggesting that Adam and Eve became knowers of good and evil after eating from the Tree of Knowledge, meaning they could distinguish between what is beneficial and what is harmful to a human being. However, they could no longer discern truth and falsehood as they could before; they could no longer understand with certainty through reason what truth should be followed and what falsehood is. According to this interpretation, The Temple of Knowledge, as mentioned by Chouchani, is the ultimate realm in which truth and falsehood are differentiated. This realm is *not* within the grasp of the descendants of Adam and Eve. This interpretation is supported by the following quote from the *Book of Proverbs*:

ה' בְּחָכְמָה יָסַד אָרֶץ, כּוֹנֵן שָׁמַיִם בִּתְבוּנָה. תְּהוֹמוֹת נִבְקָעוּ, וּשְׁחָקִים יִרְעֲפוּ טָל בְּדַעְתּוֹ.

The Lord, by **wisdom**, founded the earth; by **understanding**, He established the heavens. The depths were broken up by His **knowledge**, and the skies dropped down the dew.

The Hebrew word "בדעתו," here poorly translated into "knowledge," seems to point to something that is *beyond* human knowledge. Therefore, in his statement, Chouchani seems to point to (a) a process of human interpretation, where we realize that completing the whole puzzle of reality is a never-ending task of interpretation, and (b) a process that repeatedly brings us to the realization that some pieces just do not fit. Understanding these constraints through the Kushiya brings us, as close as possible as human beings, to the Temple of Knowledge, the lost and unachievable reality of perfect understanding. It does not get us *positively* closer to this Temple of Knowledge, but exposes our limits and, therefore, serves as a key that operates *via negativa*. Joy is possible when realizing these aspects of the quest for knowledge, and the current book dwells on them in the context of mathematics and mind, while examining the nexus between these domains.

The cognitive aspects of mathematics have been studied intensively, from the question of how children learn mathematics to the way great mathematicians think. One could hardly argue that progress in this field is even close to the progress observed in the sciences from physics to biology. One possible reason for this shortcoming is the study of the mind. The mind is still a mystery and has been a mystery since the days of Adam and Eve. The present book does not aim to resolve this mystery, only to point to some foundational aspects of mind and mathematics, and some interesting and profound nexuses between them.

Let me conclude by pointing out that, when facing deep truth, one may experience a sense of *déjà vu*. Maybe this is the reason why Plato proposed his theory of knowledge as remembering. Another explanation is that things should somehow be reflected in our minds in order to be revealed. This is what this little book is all about. Therefore, it should be read as an exercise in deep recall and no more.

Be'er Sheva, Israel   Yair Neuman

Contents

1	**The Root of Mind and Mathematics**	1
	1.1 The Root of Mind	1
	1.2 Signs of Distinction	3
	1.3 Forming Ensembles	6
	1.4 Identity and the Fundamental Law of Thought	7
	1.5 The Boundaries of the Mind	10
	1.6 Similarities and Differences	13
	1.7 Imagined Similarities	15
	1.8 From the Primary Distinction to Natural Intelligence and Mathematics	17
	References	19
2	**Equality, Similarity, and Transformations**	21
	2.1 Forms of Reasoning	21
	2.2 Play X Again, Sam	24
	2.3 What Is the Kernel of Our Dictionary?	27
	2.4 Repetition, Identity, and Mathematics	29
	2.5 Naming and Imagination	31
	2.6 Back to Repetition and Identity	32
	2.7 Freud: "Remembering, Repeating and Working-Through"	34
	2.8 Repetition in Poetry	36
	References	39
3	**Mind and Mathematics in an Event-Centered Approach**	41
	3.1 On Irreducibility	41
	3.2 In Embodiment We Trust	43
	3.3 Back to Signs	45
	3.4 Variables in Mathematics and Language	46
	3.5 Mind on the Fly	48
	3.6 Mind as Matrix and Transformation	49
	References	57

4   Symmetry, the Unconscious, and Imagination   59
    4.1   From Dada to Freud   59
    4.2   Mathematical Imagination and the Unc   65
    4.3   Condensing God and Father   66
    4.4   Repetition, Symmetrization, and Dimensionality Reduction   69
    References   73

5   Imagination, Mathematics, and Mysticism   75
    5.1   Elliot's River   75
    5.2   "What Men Choose to Forget"   77
    5.3   Negation and the Repression of Zero   81
    5.4   The Imaginary Number as a Kushiya   81
    5.5   Imagination, Mathematics, and Mysticism   88
    References   92

6   Epilogue   93

Author Index   95

Subject Index   97

About the Author

Yair Neuman is a polymath drawing on diverse disciplines to address various challenges. He published numerous papers and academic books and was a Visiting Professor at MIT, University of Toronto, University of Oxford, and Weizmann Institute of Science. His books include *Mathematical Structures of Natural Intelligence* (Springer) and *Conceptual Mathematics and Literature* (Brill).

Chapter 1
The Root of Mind and Mathematics

Abstract The root of the mind is the primary distinction, and the mind unfolds through similarities, differences, similarities of differences, and differences of similarities. We will learn about these ideas and about the mind as a sign-mediated system. These ideas and their manifestations in the mind and mathematics are discussed and illustrated through various examples, from the similarity of a sausage and a dachshund to Boole's fundamental law of thought.

1.1 The Root of Mind

The root of the mind is the primary *distinction* [1], where a void is separated into a dualistic world. This primary distinction echoes in various other distinctions: light and darkness, good and bad, God and Satan, and self and nonself are all oppositions formed when we draw a distinction. Therefore, the root of the mind is a distinction, a difference. For instance, a distinction between *self* and *nonself* is a minimal requirement for existence, and it is expressed even by "simple" structures of life, such as the bacterium. To avoid the invasion of viruses, bacteria use their immune system to recognize and attack the invaders. The bacteria must differentiate between itself and nonself to recognize and attack invaders while avoiding attacking itself (i.e., inducing autoimmunity). The bacteria must have a mind to support this distinction. In fact, actively expressing the distinction between self and nonself indicates a mind in action. In other words, the existence of living systems is mind-based [2]. A stone, a grain of sand, or a drop of water does not maintain its identity by *actively* drawing a distinction.

The mind of the bacteria is not the same mind we have, but a different kind of mind. How do we know it is a mind rather than the expression of a simple mechanical device? From the fact that the bacteria's immune system can make a mistake and attack itself. Whenever there is a mistake, there is a failure in judgment; whenever there is judgment, there is a mind. The challenge of differentiating between self and nonself is far from trivial with regard to the mechanisms involved. For our discussion, it is just another example showing that the root of all living systems, as expressed by numerous and various mechanisms, is distinction. This book opens by discussing the primary distinction from the most abstract and general perspective possible.

As argued by [1, p. xxix]: "a universe comes into being when a space is severed or taken apart." This idea was expressed long ago in the Book of Genesis:

וַיַּרְא אֱלֹהִים אֶת הָאוֹר כִּי טוֹב וַיַּבְדֵּל אֱלֹהִים בֵּין הָאוֹר וּבֵין הַחֹשֶׁךְ.

And God saw the light, that it was good, and God *divided* between the light and the darkness. (my emphasis)

According to the story, God distinguished between light and darkness and continued with distinctions to create the world. We should avoid a naïve and literal reading of this text portraying God as a *demiurge*. As explained in [3], the Bible is neither a science book nor a history book. Neither is it an engineering protocol explaining how the world was created. For the true believer, creation is beyond the limit of his understanding. This realization entails some interesting consequences. For example, I explain to my students that the ultimate rationalist must be a mystic and that the ultimate mystic must be a rationalist. The reason is that the rationalist should understand that his faculty of reason has a boundary. If there is a boundary, he must conclude that something exists beyond it. It is acknowledged that there is something beyond the realm of reason. Complementarily, the mystic understands that something exists beyond the boundary of the mind. Therefore, he must conclude that rational thought is the only thing that can portray the boundary and that he must be a rationalist to maintain it.

Adopting this approach, the only thing we can get from the biblical story of Genesis is that, metaphorically speaking, a world comes into being for human and nonhuman organisms alike when distinctions are made. Again, the biblical story tells us nothing about how God created His world, as the true believer knows it is beyond her ken. To grasp this point, imagine trying to explain to an ant the meaning of a complex number or to a human being the phenomenological experience of an ant encountering a drop of ice cream. There are things that, in a deep sense, lie beyond our understanding.

It seems that we cannot even imagine the paradoxical notion of a distinction in a world of unity, and if we have no distinction, then we have no world, either internal or external. Therefore, the primary distinction is the root of the mind. This is a fascinating idea. Teaching students in brain sciences for 7 years, I have met intelligent students who can use software to model brain signals, know about linear algebra and Chomsky, and so on, yet none of them can answer the question of the mind's root.

A distinction, suggests Spencer-Brown [1], forms a boundary separating sides that can be indicated or named. In other words, at the most basic level, a distinction forms a dualistic universe where a boundary separates two parts that can be named or indicated, such as when we call them inside and outside, or 0 and 1. The inside is the name for the unmarked state, and the outside is the name for the marked state. These states can be represented through a binary value system (i.e., 1 and 0), and any complex form involving distinctions can also be represented in binary terms. I find this idea to be fascinating. Through the primary distinction, the world can be modeled using the binary terms 1 and 0.

Spencer-Brown [1] adds layers to our understanding of the distinction. First, a *boundary* is formed. In fact, we can conceptualize the boundary as the distinction, but these nuances are less relevant than the idea that a distinction is accompanied by a boundary. Second, the two sides represent a minimally conceived binary reality formed through oppositions. Third, the sides can be named, indicated, or signified. This means that a distinction invites *naming*. When we distinguish between good and bad, we can name them as such in a way that loads the "sides" of a distinction with values.

Recursively applying distinctions to the first distinction, our world is expanded. However, we may also shrink our world back into the most basic distinction and beyond, to the unmarked state, to the void from which all began. The basic distinction can, therefore, be theoretically extended to include the totality of our world, to what will later be described as the *universe*, or shrunk to the unmarked state from which it has evolved, to what would later be described as *nothing*. The universe and nothing are two signs marking the boundaries of our mind.

But what is the meaning of the most basic distinction? Is it a thing? An operation? Moreover, if someone or something is forming a boundary, this someone must be a distinction in itself, and therefore, a distinction exists before forming a boundary. In other words, a distinction is formed by someone described as an *"observer,*[1]*"* but a distinction cannot be formed unless the observer exists as a distinction. This paradox is one of many we encounter when approaching foundational issues. Do you remember Monsieur Chouchani? Here is the first Kushiya that we encounter in this book. It may be constructive to think about these paradoxes not as simple problems that we should solve but as signs indicating that we have approached the uncharted territory of the mind. To further understand distinction as the root of cognition, let us turn to the seminal work of George Boole [4]. The reader should already have noticed my appreciation of wisdom, even if it did not appear in one of the recent volumes of Nature. Therefore, paraphrasing Newton, the book repeatedly jumps on giants' shoulders, even if they lived long before the era of scientific TikTok.

1.2   Signs of Distinction

Boole [4, p. 24], in his *Investigation into the Laws of Thought*, accepts the idea that "language is an instrument of human reason, and not merely the expression of thought." What does this mean? It means that thought, mind, or cognition, whatever you like to call it, is *mediated* through signs, where a sign is an *"arbitrary mark,* having a fixed *interpretation*" [ibid., my emphasis]. This is a nontrivial idea that

---

[1] See von Foerster: "that means the properties which are thought to reside in things turn out to be properties of the observer."

suggests our mind is what it is because we think through signs. Can we think without signs? Possibly, but when discussing complex realms such as mathematics, one cannot even imagine it existing without signs. Let me explain the idea of the sign by analyzing three of its properties.

A sign, such as a word, is a mark of distinction. In itself, the sign is a distinction. Therefore, the basic nature of the sign is *self-referential*. As explained by Kauffman [5]: "The mark is seen to make a distinction in the space in which it is written, and so can be seen, through this distinction, to refer to itself." The *self-reference* of a sign is a mystery that cannot be resolved through an act of reduction into a simpler form. Here, we encounter another Kushiya, one of many that pop up.

Being a mark turns the sign into a distinction of a distinction. This means that the sign is a representation of a distinction. For example, the sign "men" represents the class of "men." The term class is used by Boole to represent a "collection of individuals," even a single individual, or, let me add, "no individual," such as in the case of an empty set.

A simple way to illustrate this point is to use the idea of *mapping* in *category theory* [6]. A category is a mathematical construct that is comprised of a collection of things called objects, a collection of things called arrows, and operations assigning to each arrow an object from which the arrow originates (i.e., the domain) and an object to which it is targeted (i.e., the codomain). Having two sets (or classes in Boole's terms), A and B, an arrow (i.e., mapping or morphism) f from A to B means that for each element (i.e., individual) x in A, there is *exactly* one element y in B. The *value* of f at x is its element in B.

The sign is the name we give to a collection of individuals. However, the sign is actually the *value* of the function associating the domain of things with the codomain of signs. For example, if we have the signs "cat" and "bird" as two signs representing two different classes of individuals, then the mapping function assigns a mark to each collection of animals. We can see that the sign functions both as a thing and as a process.

This idea of the sign is remarkable for several reasons. The first reason is that the sign is a mark of a collective, later to be described by Boole as an *aggregate*. This means that the sign is a mark that always deals with some form of *abstraction*. It points to an abstract thing even when it is used to point to concrete objects. Moreover, it is the name of a class, even if the class has no individuals in it or even if it includes a single individual. The sign is not a thing's name but the name we give to the *value of a function*. An infant pointing at the cat and saying "cat" is actually naming the value of the morphism through which she assigns an object, which is a sign, to an object, which is a thing signifying a set of objects (i.e., her mental concept). This is a nontrivial achievement for an infant, as it encapsulates all higher-level forms of abstraction that we later observe in mathematics. We will ponder upon these ideas later.

Boole further suggests that the sign also has an operational sense. For instance [ibid., p. 44]: "The word 'men' implies the operation of selecting in thought from its subject, the universe, all men." The sign, as a mark, does not simply name or point to a certain class, but functionally and actively *selects* it from a universe of things.

It is an *attentive act* in which our mind illuminates a certain aspect of our universe, enacting some of its aspects while ignoring others. As such, through signs, our mind is not a passive observer of reality but an active participant in reality. Again, I conceive it as a remarkable idea. Sometimes, the sign is naively portrayed as just a name. However, Boole explains that indication is much more than "naming." It is an activity in which the mind is attuned to certain aspects of a world, whether real or imagined, while inevitably ignoring others. Enacting certain aspects of the world reminds us of Bohm's proposal [7] to consider our measurement of the world as actively coupled with the measurable world.

Boole further differentiates between two types of signs: signs of *things* and signs of *operations*. However, if the sign is both the "name" of a class and the operation of constraining our observation to certain classes, then it has a *Janus* face. On the surface, the sign describes the invariance of a thing. A cigar is just a cigar, to paraphrase Freud's famous statement. However, below the surface, it does not correspond to a noun or a thing but to an activity that results in conceived invariance. This idea has a radical constructivist flavor: the world that seems to be passively reflected in our mind, in the same way that a mirror reflects our face, is actually the construction of our mind *actively* adhering to certain aspects of the world while ignoring others. The world is not passively given but is actively constructed. One should not confuse this idea with naive postmodernist approaches that describe our world as *arbitrarily* constructed according to some whimsical narratives or narratives of dominance and hegemony. The *woke* culture, which is unfortunately prevalent in some areas of the academic world, has raised this approach to a level of decadence. However, a postmodernist bacterium arbitrarily representing its self and nonself would have ceased to exist. The bacteria's mind actively constructs a world that corresponds to the niche in which it has evolved. Reality is evident wherever our fantasies hit a wall, and despite its complexity, realities firmly exist for all living things. The reality facing the bacteria differs from that facing a rat, but it is a firm reality for each species. The lesson we should learn from the constructivist aspect of cognition previously presented is mainly a lesson in modesty. Those who strive for a simple and complete understanding of our world or adopt a single and narrow perspective while ignoring different realities should understand that the first limit line of our understanding is embedded in the basic root of our mind.

The second aspect of the sign is that it is *arbitrary*. It is arbitrary because its "material" cloth is irrelevant to understanding its value. The value of the sign is determined only through the mapping function, as explained before. Different languages use different signs to designate the same things. The word "cat" in English marks the class of feline creatures, while the same class can be represented by the word חתול in Hebrew. The sign's meaning is not found in the sign itself but in its function: mapping from a domain of classes to the codomain of marks. Interestingly, the relation (i.e., the mapping) *precedes* the things it associates.

The arbitrariness of the sign and its mapping-based sense allows enormous freedom for the mind. Think, for example, about the *imaginary number*. Did it exist before it was imagined? And if so, in what sense? For another example, think about the word God and what it signifies. Leibowitz (1903–1994), one of the great modern

Jewish thinkers and a scientist with degrees in chemistry and medicine, explained [3] that the unique perspective of Jewish monotheism is that God is not the world and cannot be identified with the world or any aspect of it. Idolatry identifies God with concrete references: the world, nature, mountains, or statues. Adopting Leibowitz's approach, the sign "God" is meaningless for any theological exploration. It does not point to anything in our world, and the name itself tells us nothing about the thing that it signifies. The consequences of this approach are radical.

The third aspect of the sign is that it must have a fixed interpretation. This doesn't mean that meaning does not change over time or across different contexts. Having a fixed interpretation is an idea that can be explained again by reference to category theory. The *value* of a sign is the element to which it is assigned in a codomain. Having a fixed interpretation means that the sign has a value and *only one value* in a given context of use. A sign may be characterized by *polysemy*, having different senses in different contexts, but it must have only one interpretation at a time, one value defined by a mapping function. Interpretation, meaning, and value are the same and may be conceived in terms of *mapping*.

In sum, the sign is (1) a mark, (2) arbitrary, and (3) has a fixed interpretation. As proposed by Boole, we think through signs; mathematics cannot exist without signs, and therefore, understanding mathematics is understanding human thought as a semiotic (i.e., sign-mediated) system.

Boole further argued that [4, p. 21]: "All the operations of Language, as an instrument of reasoning, may be conducted by a system of signs" composed of "symbols representing things as subjects of our conceptions," signs of operations (e.g., +), and the sign of identity "=." In other words, a world of distinctions can be indicated through three types of signs indicating things, operations, and identity. For example, the equality sign "=" was first introduced by the Welsh mathematician Robert Recorde (1557), who explained it as a sign replacing the longer expression "is equal to." Historically, this sign was formed to replace an idea that existed in thoughts and was expressed through a longer expression in natural language: x is equal to y.

1.3   Forming Ensembles

According to Boole, the combination of signs and operations is the basic form of language and, therefore, a gate for understanding the laws of human thought. For example, Boole suggests [ibid., p. 31] that combining two signs represents the whole of "that class of objects." If x is the sign of "men" and y is the sign of "tall," then their combination xy represents the class of tall men.

The combination of classes can be discussed in the most general sense and does not have to be limited to certain classes of objects, such as sets, or to combination through specific operations, such as *union*. Category theory proposes a specific concept for discussing this general notion of sum. It is the concept of the *coproduct*. Let me explain summing in the most general sense. When combining the class of "tall" with the class of "men," we get a new class of "tall men." In the language of set theory, the new set of tall men is the intersection of two sets: the set of men and the set of tall things. Some men are tall, and others are not. Some tall things are men, and others are not. Forming the set of tall men is identifying the intersection of the two sets. A similar logic appears when we merge two sets through their union. If we have the set of dogs {poodle, German shepherd, dachshund} and the set of food {bread, cucumber, cheese}, then their union is the set {poodle, German shepherd, dachshund, bread, cucumber, cheese}. However, union is not the only way to form a composition or ensemble. When we form the word composite "hotdog," we cannot form the meaning of the whole word through the sum of the semantic components of "hot" and "dog," nor through their intersection. A word composite is not a simple aggregate or intersection but an *ensemble*, a concept discussed in physics by Bohm [7]. For an ensemble, we must use a more general concept, such as the coproduct, that describes how different things are put into the same mental basket to form a collective (i.e., a Gestalt) different from the sum of its parts.

To better understand the composite word as an ensemble, we may consider the word "hotdog" as composed of two vectors: one representing the meaning of "hot" and the other representing the meaning of "dog." Each vector is an array of numbers representing the extent to which a set of predefined words appears together with our target word, whether "hot" or "dog." This form of representation captures the meaning of the word through the semantic field in which it is located, and it is the form of representation underlying the most sophisticated technology of *Large Language Models* (LLMs). In this context, understanding the meaning of "hotdog" is not a simple arithmetic sum of the vectors composing hotdog, but a transformation and combination of the two vector spaces representing "hot" and "dog" into a whole (i.e., ensemble) that is different from the sum of its parts. The meaning of "hotdog" is not determined by simply adding two semantic spaces. LLMs explain that the meaning is formed by adjusting each word's vector to the other word's vector to form the composite word "hotdog." Combining parts into a whole is a process that may have different expressions, some of them far more complex than the union of sets. We should not think of this operation from one simple perspective.

1.4 Identity and the Fundamental Law of Thought

The combination of two signs (e.g., xy) "expresses the whole of that class of objects to which the names or qualities represented by *x* and *y* are together applicable" [4, p. 22]. Therefore, $xy = x$ only if $x = y$. Boole further concludes that $x^2 = x$. One may wonder how algebra and multiplication got into the picture. Boole explains that x^2 expresses the idea that "a particular succession of mental operations is *a thing in itself*" [ibid., my emphasis]. This is an important point as it shows us that identity can be derived from the operation of multiplication. Drawing on algebra, Boole explains that $x \cdot x$ can be represented as x^2. This notation expresses "the succession of mental operations." In other words, repeating the same operation would result in the same class. This idea is equivalent to Spencer-Brown's first axiom, described as the "law of calling" [1, p. 1]: "The value of a call made again is the value of a call."

Why use multiplication and not addition to prove that something remains constant? As young children, we are taught that multiplication is just an addition in disguise. So why multiplication? Instead of asking what x is, such that x • x = x, we could have asked what x is, such that x + x = x. The answer to the second question is trivial. It is 0. According to Peano's axiomatic foundations of arithmetic, adding nothing to nothing changes nothing. Using multiplication and the variable x, we can move beyond the strange domain of 0 and into a dualistic realm of the primary distinction, where the solution to the equation $x^2 = x$ can be x = 1 or x = 0, the two values of the primary distinction. Regarding multiplication as an addition, it is better to think of multiplication as *scaling*, stretching, or shrinking an object by some *scaling factor*. Figure 1.1 visualizes multiplication as scaling.

Multiplication as scaling can help us understand how the identity of 1 and 0 remains unchanged by multiplying the number by itself. Later, I plan to return to the idea of multiplication and scaling and, in the concluding chapter (Chap. 5), to Boole's algebraic idea of identity.

Nothing can be said unless there is something stable in our reality, and stability is epitomized in identity. Unless x = x, there is no x. Therefore, it is important to understand that Boole is telling us something interesting about identity and time. *Identity* is formed only in *time* and through *repetition,* which can be imagined as a linear *transformation* (scaling). As he explains [4, p. 32], "Most of the operations which we observe in nature, or perform ourselves, are of such a kind that their effect is augmented in repetitions." There is no identity outside of time and repetition. Identity, described by Boole as a foundational axiom, is, therefore, a mathematical concept grounded in the mind.

One may argue that identity can be formed through the existence of certain *identity objects*. For example, 3 • 1 = 3. This equation shows us that, with respect to natural numbers and the operation of multiplication, the identity object is always 1. There is no repetition here, at least not in the simple senses presented by Boole. However, 3 • 1 is actually equivalent to pointing again to a class of three objects. It adds nothing to the distinction we already observe. The same is true for 3 + 0 = 3. Adding nothing to the distinction is a repetition of the same distinction. Nothing changed.

Fig. 1.1 Multiplication as scaling. (Source: Wikipedia)

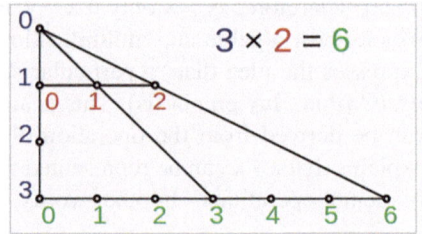

The analogy with algebra is further investigated by Boole, who explains that there are two numbers that are the solutions to the equation $x^2 = x$. These numbers are 1 and 0: $1^2 = 1$ and $0^2 = 0$. The equation $x^2 = x$ has no roots other than 1 and 0. Boole describes $x^2 = x$ as *the fundamental law of thought* and illustrates how the *principle of contradiction*, stating that "it is impossible that the same quality should both belong and not belong to the same thing" [ibid.], is derived from the basic law of thought:

$$x = x^2$$
$$x - x^2 = 0,$$
$$x(1 - x) = 0$$

The final expression explains that a thing and its complement in the universe result in nothing, meaning that the intersection of a thing and a nonthing is an empty set. For example, the class in which members are both men and not men does not exist. It does actually exist but is signified as the empty set "{ }."

Interestingly, the equation, $x(1-x) = 0$, expresses Spencer-Brown's second axiom, described as the *law of crossing*: "The value of a crossing made again is not the value of the crossing." Let me explain this law. A distinction is made, explains Spencer-Brown, by arranging a boundary so that a point on one side cannot get to the other side without crossing the boundary. We can visualize this move in Fig. 1.2.

The point crossing from the inside to the outside forms the distinction. It is x. Crossing back is represented by the second part of the above expression: (1-x). The return from the outside is necessary to form the universe composed of inside and outside. Therefore, it is an operation of *cancellation* bringing us back to nothing: $x(1-x) = 0$. In fact, Boole's first law of thought shows Spencer-Brown's second axiom to be unnecessary, as it is derived from the first.

The fundamental equation of thought, explains Boole, is of a binary nature (x is raised to the power of 2) and is satisfied when $x = 0$ or $x = 1$. This is why duality and opposition [8] are at the heart of our minds. When forming the primary distinction, the marked and the unmarked states are symmetrical and *mutually defined* with respect to each other. There is no light without darkness and no self without a non-self. This is why the mind is grounded in oppositions, and symmetry is the hallmark of the unconscious, an idea to be developed and explained in one of the following chapters.

Fig. 1.2 Crossing.
(Source: Author)

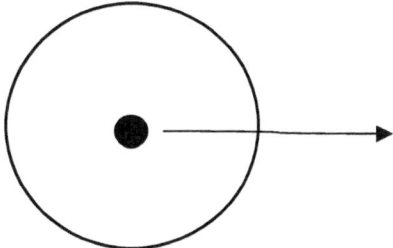

1.5   The Boundaries of the Mind

Let us return to the idea of a distinction. Consider the primary distinction. This distinction forms a binary world, separating this from that, or, according to Spencer-Brown, inside from outside, the unmarked from the marked state. Let us indicate one side of the boundary as x and the other as the whole minus x: 1–x. The value 1 signifies the whole, described by Boole as the *universe*, the collection of all things, and 1 is the sign representing the class where all things exist. There are no things beyond the universe. It is the sign indicating the upper limit of our distinctions, and it is a necessary construct for defining the basic form or the primary distinction. If a distinction is made, it forms two different things. Unless defined as the individuals of a universe, these things cannot form a distinction through which their existence is mutually defined as a *negation* of each other.

To understand this point, we may use category theory and the idea of a *terminal set*. Lawvere and Rosebrugh [9, p. 6] propose the following axiom: "There is a set 1 such that from any set A there is exactly one mapping A → 1. This unique mapping is given the same name A as the set that is its domain."

While the idea may sound abstract, it can be easily explained through WordNet,[2] a lexical database that organizes words/concepts in a hierarchy. Consider the word "cat." If you trace its inherited hypernym in WordNet, you learn that a cat is a feline, a carnivore, a mammal, and so on, until you reach the top-level category, "entity." The category of entity is defined in WordNet as "that which is perceived or known or inferred to have its own *distinct* existence (living or nonliving)" (my emphasis). So, a cat is an *entity*, a concept signifying the limit line of the taxonomy. No class is located above the class named "entity." In fact, the set of all real or imagined things can be mapped to the terminal set titled "entity." This is the sign we use to designate a universe. Nothing exists beyond it. Actually, there is. However, our minds cannot comprehend what exists beyond the boundary. Therefore, the sign "1" is like a road sign saying something like: "Danger! Abyss ahead."

The idea that only one map exists from A to 1 means that all things are assigned the same sign/name. The value of a cat, a chair, and the idea of freedom are "entities," as they are all distinctions. There is nothing we can say about them beyond that, and the terminal set is, therefore, where *all distinctions are lost*. Give yourself a minute to think about this idea. Mind and mysticism are deeply connected through the idea of the terminal set.

Lawvere and Rosebrugh [9, p. 7] further suggest the following definition: "An element of a set A is any mapping whose codomain is A and whose domain is 1." The idea is that any element in a set is defined through a mapping function originating from 1. Let's explain this idea through the following *lattice,* where x and y are the minimally distinct things comprising the primary distinction. A *lattice* is "an abstract structure ... consisting of a partially ordered set in which every pair of elements has a unique supremum (also called a least upper bound or join) and a unique

---

[2] https://wordnet.princeton.edu/

Fig. 1.3 The lattice of the
primary distinction.
(Source: Author)

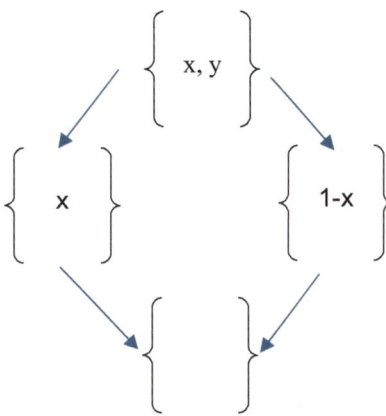

infimum (also called a greatest lower bound or meet)" [Wikipedia]. Figure 1.3
shows the lattice of the primary distinction.

The universe comprises x and y, which can also be described as x and 1-x, the
inside and outside. These elements are mutually defined as oppositions through
negation: 1 is NOT (0), and 0 is NOT (1). The existence of x and $1-x$ is established
by a mapping from the universe $\{x,1-x\}: 1 \rightarrow A$. Therefore, the elements x and $1-x$
are self-referentially defined by 1, the same universe in which they are the composing elements. Here, another Kushiya pops up unless one accepts the idea of a *recur-
sive hierarchy* as built into existence.

However, the upper limit formed by 1 is not the only limit required for understanding the meaning of the primary distinction. As explained by Boole, the universe or the whole goes hand in hand with *nothing*, signified as "0." They are the
"limits of class extension" [4, p. 47].

This nothing, the empty set or the *initial object*/set, is where the distinction
emerged from. The theoretical construct points to where no distinctions exist, the
sign we use to describe the collection of individuals that includes only one "individual": nothing. This nothing or the empty set, though, is also a distinction, the
distinction between a collection that is empty (i.e., includes no members) and other
collections that are not empty. As you can see, distinctions cannot be avoided even
when we speak about nothing. We are doomed to distinguish. At the same time, we
cannot understand how a distinction can evolve from nothing: A Kushiya.

Here, we can better understand why signs exist when we encounter complex
mind systems. To represent the system's boundaries, we require signs paradoxically
indicating what is beyond our grasp. Moreover, do you remember the idea that
$x^2 = x$? The ultimate identity of the elements composing the primary distinction is
guaranteed using x = 1 or x = 0. In both cases, the aggregate of the collection with
itself results in identity in the strongest possible sense, as the existence of 1 and 0 is
guaranteed through axioms. The boundaries of our mind are also those establishing
the only and strongest sense of identity. An axiom [9 , p. 12] states that there is a set
0; for any set A, there is exactly one mapping from 0 to A. No mapping exists from

A to 0, as 0 is empty and has no elements. For everything and nothing to exist, there must be an observer/mind making distinctions, and for distinctions to exist, we must accept the axiom that 0 is not equal to 1. But if 0 is different from 1, then this is another distinction that requires some limits! This is why we consider the axioms of the universe and nothing, as they present a limit line to our understanding. This is an important point; the limits set boundaries for understanding and interpretation.

The Jewish sages well understood this boundary when they read and interpreted the opening sentence of Genesis. The opening is as follows:

בְּרֵאשִׁית בָּרָא אֱלֹהִים אֵת הַשָּׁמַיִם וְאֵת הָאָרֶץ

Translation: In the beginning, God created the heavens and the earth.

As Hebrew is written from right to left, the sages were trying to understand why the first letter opening the bible is the Hebrew letter ב. They explained it as follows:

לָמָּה נִבְרָא הָעוֹלָם בְּב'? אֶלָּא מַה ב' זֶה סָתוּם מִכָּל צְדָדָיו וּפָתוּחַ מִלְּפָנָיו, כָּךְ אֵין לְךָ רְשׁוּת לוֹמַר מַה (לְמַטָּה, מַה לְמַעְלָה, מַה לְפָנִים, מַה לְאָחוֹר, אֶלָּא מִיּוֹם שֶׁנִּבְרָא הָעוֹלָם (וּלְהַבָּא (מדרש בראשית רבה, א, י

Translation: "Why was the world created with the letter 'Bet' (ב)? Rather, because the letter 'Bet' is closed on all its sides but open in front of it similarly, you have no permission to say what is below, what is above, what is in front, and what is behind, except from the day the world was created and onward."

The sages acknowledged the limits of interpretation and the necessity of axioms for any human inquiry, from mathematics to hermeneutics. Moreover, elsewhere, they even *warned* us about crossing this boundary of axioms:

כָּל הַמִּסְתַּכֵּל בְּאַרְבָּעָה דְּבָרִים, רָאוּי לוֹ כְּאִלּוּ לֹא בָּא לָעוֹלָם:

1.   ,מַה לְמַעְלָה

2.   ,מַה לְמַטָּה

3.   ,מַה לְפָנִים

4.   .וּמַה לְאָחוֹר

(וְכָל שֶׁלֹּא חָס עַל כְּבוֹד קוֹנוֹ, רָאוּי לוֹ שֶׁלֹּא בָּא לָעוֹלָם (משנה חגיגה ב א

Translation: "Anyone who reflects upon four things, it would have been better for him if he had not come into the world:

1. What is above,
2. what is below,
3. what is in front,
4. and what is behind. And anyone who has no regard for the honor of his Maker, it would have been better for him if he had not come into the world." (Mishnah Hagigah 2:1)

Maimonides explains this warning by saying that inquiry into what is beyond the boundary has dangerous consequences:

"תוציא אותו זו המחשבה אל השגעון ותמהון הלב."

Translation: "Such contemplation leads to madness and perplexity of the heart."

By inquiring into the foundations of mind and math, we understand the limit line of our understanding.

1.6 Similarities and Differences

The Universe is formed by "collecting" parts into a whole. Boole [4, p. 32] describes signs "whereby we collect parts into a whole or separate a whole into its parts." The sign "and," signified by Boole as "+," forms a whole. This sign is not used in its limited arithmetic meaning. Indeed, Boole defines the sign "+" as the operation *aggregating parts into a whole*. Saying that a cat and a human being are mammals forms a higher-level class in the lattice of our taxonomy, which comprises two parts: cat and human being. Each part is a set indicated by a sign. It is a marked distinction.

The complementary operation to + is signified as −. This sign indicates *breaking a whole into its parts*. The concept of a mammal is meaningless without the subsets composing it, and a thing cannot be formed without aggregating some lower-level things into a whole.

One should notice that + is the mental operation expressing *similarity*; if x, y, and z are things aggregated into a whole, they are similar. Similarity can be expressed in different forms. Aristotle already realized in his *Rhetoric* that associations may be formed through the similarity (homoiotēs) of qualities. For instance, the association between sausage and dachshund is formed through the similarity of their visual structure (Fig. 1.4).

The sausage is not identical to the dachshund. In fact, only a madman would mistake a dachshund for a sausage and try to eat it. Confusing similarity with identity is a fallacy repeatedly performed by the unconscious part of our mind. In fact, madness may be expressed as confusing similarity with identity.

Grouping things by similarity or contiguity in time/space may explain how we aggregate individuals into wholes. Complementarily, the reverse operation "−" is

Fig. 1.4 A sausage and dachshund. (This image was created with the assistance of DALL·E 2)

responsible for creating differences. For example, in the dualistic cosmology of *Manichaeism*, there is no light without darkness, and vice versa. The difference between the world of light and the material world of darkness is possible only through the subtraction of light from a universe where light and darkness coexist.

The importance of similarity/difference, deeply associated with Boole's two operators, cannot be underestimated. As argued by David Bohm [7], our minds are grounded in similarities, differences, similar differences, and different similarities.

For instance, consider a distinction we name "mother." This distinction is possible only through the class named mother. Is it possible to have the idea of a mother without a sign indicating the thing called a mother? Let's elaborate on this point through a concrete example.

My granddaughters enjoyed a book showing pictures of mothers and their offspring. Each page presents a picture of a mother and her offspring and asks the infant to identify who the mother is. The pictures look like the images in Fig. 1.5.

Learning the idea of "mother" involves finding similarities between differences. The human mother is different from the chicken and the mother cat. The human baby is different from the chick and the kitten. However, each picture depicts a relationship: the human is the MOTHER of the baby, the chicken is the MOTHER of the chick, and the cat is the MOTHER of the kitten. Despite their differences in appearance, all "mothers" are aggregated into the same class under the same title. By aggregating them into the same class, they become similar. Their similarity is not given but constructed by the mind. Our mind identifies similarities and differences. Despite their differences, all "babies" are aggregated into the same "baby" class. Again, the different babies are conceived similarly by aggregating them into the same class. The "babies" are not similar in any real way, expressed in similar perceived features. They are metaphorically "similarized" through the activity of

Fig. 1.5 Mothers and their offspring. (This image was created with the assistance of DALL·E 2)

the mind and its capacity to imagine the possibility of grouping a chick, a kitten, and a human baby under the same roof. The above example shows how imagination, in the Whiteheadian sense to be discussed later, appears at an early age. "Seeing" beyond the sensorimotor differences indicates an incredibly sophisticated process of the mind. Only through imagination and signs can the infant go beyond the perceived reality.

The general conclusion is that we may conceive different objects as similar by grouping them into the same class. This is a highly important point for understanding mathematical objects, as will be explained shortly. In this case, the similarity is not formed through a low-level similarity of *association-by-sense* [10]. The objects are similar through a high-level function, associating them with another class of different things. The mapping function forms the similarity within each class and between different objects. The interplay of similarity and difference is a powerful process underlying the mind.

1.7   Imagined Similarities

We may better understand this construction using the idea of a *functor*. A functor is "a transformation from one category into another that preserves the categorical structure of its source" [6, p. 194]. Let's describe two categories. In the first, we have as objects mothers assigned to offspring. The morphism f assigns to each mother her offspring. For example, the mother is assigned to her baby, f: m → b, and the cat to its kitten, j: c → k. The morphism from f to j forms the similarity between the different objects. Similarizing the relation between mother and baby (m:b) with the relation between the cat and the kitten (c:k) forms an analogy (m:b::c:k) underlying the general concept of a mother: The mother is to the baby the same as the cat is to her kitten. We can better understand the idea if we think about it by forming the idea of a number and performing analogical reasoning.

In a deep sense, learning the idea of a mother is the same as learning an analogy or the concept of a number. Let me explain. When we learn to count, we learn to signify through numbers. What is this thing we call number? Imagine a child counting to three: "One, two, three." He first counts three of his fingers, then three apples, and then three of his candies. How can it be that these three different sets all gain the same sign, which is "3"? Russell [11] explains that a number, such as "3," is the sign of all collections containing three elements. But isn't that a circular argument? Russell explained that the cardinality of this collection (i.e., its size) can be established if there is a one-to-one mapping between its elements (i.e., a bijection). The number of a class, Russell proposed, is the "class of all those classes that are similar to it" [ibid., p. 18]. The similarity is a similarity of differences. Although the classes are comprised of different things grouped in each separate set, they are assembled into a higher-level class named "3" through a mapping function. The number 3, arbitrary as a specific mark, seems necessary to identify the sets' similarity. It is a distinction that exists at a higher level of abstraction and a distinction that forms a new concept "3." We create the similarity between differences through this sign, which is a mapping function. With no signs at our disposal, we would have been doomed to live in a world of relatively simple distinctions grounded and constrained by our immediate sensorimotor experience.

Russell's idea draws on basic distinctions that can be easily represented by the notation proposed by Spencer-Brown [1]. Here is the representation of the numbers 0 to 3 using ר as the sign of the marked state.

> 0 = the unmarked state
> 1 = ר
> 2 = ר ר
> 3 = ר ר ר.

However, numbers are not just distinctions. They are *signs* pointing to a deep similarity (i.e., analogy) of distinctions beyond the particularities in which they are expressed (e.g., three apples vs. three kittens). The logic of numbers is expressed in the sign-mediated interplay of similarity and difference across different levels of abstraction (i.e., ensembles). When we understand that the sign of a number is an arbitrary mark creating a similarity of differences, we may start playing with it to create other distinctions.

What about *analogies*? Consider the analogy one should identify:

The hand is to the glove as the foot is to the _____?

The answer choices could include:
(a) nail (b) shoe (c) hat (d) scarf.

The correct answer would be "shoe," since, just as a hand fits into a glove, a foot fits into a shoe. Notice that a hand is different from a foot, and a glove is different from a shoe. However, the analogy establishes a similarity of differences.

In building a computational model for solving analogies, Turney [12] differentiated between domain similarity and functional (or relational) similarity. Hand and glove have domain similarity, as do foot and shoe. They are entities related by their coexistence in the same semantic field. To solve an analogy, Turney [12] proposes a computational approach that maximizes domain similarity between certain elements and functional similarity between others. A hand and a foot have a certain functional similarity, as do a foot and a shoe. Let us represent the analogy as follows:

$$H : G :: F : ?$$

To identify the missing part of the analogy and solve it, we should look for the answer that maximizes domain similarity with the foot, minimizes domain similarity with the hand, and maximizes functional similarity with the glove. The answer is "shoe." Solving the analogy involves the same process as forming the idea of a number or the idea of a mother. In all cases, we identify a similarity of differences where the similarity is formed between relations.

Now, we can return to the example of the mother. Being a mother is the name of a class of individuals. This class is formed through a similarity of relation, or what is described in [12] as functional similarity. This similarity is a similarity of relation: a cat gives birth to a kitten in the same way that a mother gives birth to a baby. The similarity of relation is the one aggregating different things into the same class.

1.8   From the Primary Distinction to Natural Intelligence and Mathematics

In 1899, Herbert Spencer Jennings published "The Psychology of a Protozoa" [13]. The protozoan is a single-celled, cigar-shaped creature that was intensively studied by Jennings. He was struck by the complex behavior of this creature and asked [ibid., p. 506]: "Is it possible by a closer analysis of the phenomena to simplify this complex psychology which seems forced upon us by the observed facts?" Looking further into this question, Jennings answered it negatively [ibid., p. 515]:

> While we cannot deny that Paramecium is an organism, this fact shows the machine-like nature of its activities. An animal that learns nothing, that exercises no choice in any respect, that is attracted by nothing and repelled by nothing, that reacts entirely without reference to the position of external objects, that has but one reaction for the most varied stimuli, can hardly be said to have made the first step in the evolution of mind, and we are not compelled to assume consciousness or intelligence in any form to explain its activities.

Jennings was wrong. It took some years to realize that even bacteria learn, but on a different level of organization, as explained in [14, emphasis mine]:

> Among the many bacterial defense mechanisms against phages, CRISPR-Cas is unique in its ability to *learn* from past infections by storing pieces of phage DNA (called spacers) in its own genome to neutralize future infections.

A system that adapts is a system that learns; to learn, one must have a memory. In our discussion, the mechanism is less important than the phenomenon: even the simplest life forms can make distinctions, such as the one between self and foreigners. So, what is the difference between bacteria and the mind of a human being? Luria and Vygotsky [15] proposed long ago that the human mind should be studied via three interwoven threads: evolutionary-biological, developmental, and cultural. The first thread asks us to understand the human mind in the context of the great chain of beings of which we are a part. In this context, we realize that even the simplest life forms may have "basal cognition," which is characterized, among other things, by the ability to discriminate [16]. We can see that the ability to make distinctions is a basic cognitive ability. Different beings form different kinds of distinctions, some of which are more sophisticated than others. However, by examining our mind from the broader perspective of living systems, we realize that distinction is the most basic form of the mind, whether in the mind of Einstein or the mind of bacteria.

The developmental thread suggests that something happens to us as we mature. Our world is potentially enriched by extending our distinctions and moving up the ladder of abstraction. When my granddaughters first learned the idea of "mother," they used it to describe the relation between big and small entities that have a similar appearance. A big chair has been described as the mother of a small chair, and a big shoe could have been described as the mother of a small shoe. It took them some time to understand that being a mother is not about size. Motherhood is not exhausted by size but by a subtle, more abstract relation that cannot be trivially grounded in any sensorimotor experience. A mother gives birth to offspring, while the "mother chair" doesn't deliver its "baby chair." Children take time to learn it.

In contrast to primitive life forms, human beings present a relatively longer development phase, closely related to the third thread: culture. Culture is the name for a collective sign-based activity through which our minds are mediated. Learning math and complex mathematical ideas is not a natural development phase. Some cultures have never developed mathematics in the same way as others. The woke call for the "decolonialization of mathematics" ignores this fact. Mathematics is learned in certain *cultures* at the appropriate developmental phase. You cannot learn algebra before learning to count, and you cannot learn to count before making some distinctions and learning the symbolic activity that associates numbers with classes. This symbolic activity is learned through cultural resources to which no one has copyright. Therefore, mathematics does not belong to certain groups of people and cannot be "colonized" or "decolonized." Mathematical thinking is grounded in the basic notion of distinction and in our ability to identify similarities, differences, similar differences, and differences of similarities through a gradual developmental and acquired process of semiotic mediation.

Points to Note

- **The root of the mind**: the primary distinction.
- **The dualistic world**: formed by opposition (1 and 0).
- **The sign-mediated realm**: signification moves us into a realm where signs support abstraction and imagination.
- **The sign as value and attentive act**: signs are both values of functions and acts of attention.
- **Identity formation**: identity is formed over time through repetition and transformation.
- **The fundamental law of thought**: $x^2 = x$.
- **The interplay of similarity and difference**: characterizes expressions of the mind.
- **Learning numbers**: the concept of numbers is similar to analogies and names.

References

1. Spencer-Brown, G.: Laws of Form. Cognizer Co, Portland (1994)
2. Neuman, Y.: Mathematical Structures of Natural Intelligence. Springer, New York (2017)
3. Leibowitz, Y.: Seven Years of Discourses on the Weekly Torah Reading. Keter, Jerusalem (2000) (in Hebrew)
4. Boole, G.: An Investigation of the Laws of Thought. Dover Publications, New York (1854)
5. Kauffman, L.H.: Laws of form: a survey of ideas. In: Kauffman, L.H., et al. (eds.) Laws of Form—A Fiftieth Anniversary, pp. 1–88. World Scientific Pub Co, Singapore (2023)
6. Goldblatt, R.: Topoi. North-Holland, Amsterdam (1984)
7. Bohm, D.: On Creativity. Routledge, London (2004)
8. Danesi, M.: Opposition and semiosis. In: Pelkey, J. (ed.) Bloomsbury Semiotics. History and Semiosis, pp. 215–240. Bloomsbury Publishing, New York (2022). https://doi.org/10.5040/9781350139312.ch-10
9. Lawvere, F.W., Rosebrugh, R.: Sets for Mathematics. Cambridge University Press, Cambridge (2003)
10. Danesi, M.: Metaphorical "networks" and verbal communication: a semiotic perspective of human discourse. Σημειωτκή-Sign Syst. Stud. **31**(2), 341–364 (2003)
11. Russell, B.: Introduction to Mathematical Philosophy. Routledge, London (1993)
12. Turney, P.D.: Domain and function: a dual-space model of semantic relations and compositions. J. Artif. Intell. Res. **44**, 533–585 (2012)
13. Jennings, H.S.: The psychology of a protozoan. Am. J. Psychol. **10**, 503–515 (1899)
14. Bonsma-Fisher, M., Soutière, D., Goyal, S.: How adaptive immunity constrains the composition and fate of large bacterial populations. Proc. Natl. Acad. Sci. **115**(32), E7462–E7468 (2018)
15. Luria, A.R., Vygotsky, L.S.: Ape, Primitive Man, and Child: Essays in the History of Behavior. CRC Press, Boca Raton FL (1992)
16. Lyon, P., Keijzer, F., Arendt, D., Levin, M.: Reframing cognition: getting down to biological basics. Philos. Trans. R. Soc. B. **376**(1820), 20190750 (2021)

Chapter 2
Equality, Similarity, and Transformations

Abstract The equality sign (i.e., "="), the idea of identity, and repetition are discussed. I explain how repetition stands at the heart of identity and point out the similarity of repetition in mathematics, psychology, and the arts. By discussing iterated function systems, fixed points, and invariant sets, the reader will learn that the *equality sign in mathematics indicates a symmetry expressed under certain transformations.* This chapter takes us on a journey from mathematics to psychoanalysis and poetry.

2.1   Forms of Reasoning

Poincaré [1] described mathematics as "the art of giving the same name to different things." However, when a child learns to name different animals using the sign "animal," he also gives the same name to different things. Is he doing mathematics? The child expresses the mind's general ability to abstract by identifying similarities between different collections of things. However, mathematics expresses a unique form of abstraction. Mathematics is the art of pointing out deep similarities between *mathematical objects* at increasingly higher levels of abstraction. One prominent mathematical object is the number. We first learn to count using the same name—the number—to describe sets of different things characterized by the same cardinality. The idea of a number, as discussed in the previous chapter, is just one example illustrating Poincaré's insightful observation. A number (i.e., natural number) is a sign we use to mark the similarity between collections that may include different objects. Some collections may include chairs; others may include ravens or cucumbers. However, we indicate their similarity whenever we conclude that all these sets contain the same number of things. The sets are similar to each other *despite* the different things they include. The result is surprisingly the same whether one counts five apples, five cats, or five frogs: it is 5. I use the word "surprisingly" because the "Aha!" phenomenon is not an experience exclusive to great mathematicians. It is the experience of each and every child who suddenly realizes the beauty of a deep similarity existing beneath a surface of differences. This is the beauty of recognizing a

Y. Neuman, *Mindmatics*, Mathematics in Mind,
https://doi.org/10.1007/978-3-031-74955-1_2

structure, whether in a sunflower or a deep mathematical representation, as presented in *The Bridges of Königsberg*. Most of the time, this beauty is masked or missed when we study mathematics in primary or high school. The tiresome memorization and problem-solving of artificial problems I experienced as a school student distract our minds from this beauty. Maybe school mathematics never aimed to show us the beauty of structures and reasoning, only to train low-level accountants for the naval office or other branches of modern bureaucracy.

Mathematics, though, is not only about giving the same name to different things. Mathematics relies heavily on a unique form of reasoning. As argued by Whitehead [2, p. vi, emphasis mine]:

> Mathematics in its widest signification is the development of all types of formal, necessary, deductive *reasoning*. The reasoning is formal in the sense that the meaning of propositions forms no part of the investigation. The sole concern of mathematics is the inference of propositions from propositions.

For Whitehead, at least according to the above statement, mathematics is about reasoning from one proposition to another, regardless of the proposition's meaning. What is important is the *form* of reasoning and nothing else. When I studied logic as an undergraduate student, I learned a form of syllogism that can be illustrated as follows:

If
All squirrels are mathematicians
and
Johnny is a squirrel
Then
Johnny is a mathematician

This chain of reasoning is valid regardless of the propositions' truth values. The proposition "All squirrels are mathematicians" is not only false but also meaningless. Human beings alone can be described as mathematicians. The proposition "Johnny is a squirrel" may be accepted as true with respect to a pet squirrel called "Johnny." Therefore, we have a meaningless premise, a second premise that can be meaningful and valid, and a conclusion necessarily derived from the premises. The conclusion is valid despite being false and meaningless. Therefore, the validity of a syllogism and the truth of the premises or the conclusions are not the same. For some students, this process of deductive inference seems counterintuitive because they conflate the truth value or meaning of propositions with logic. Logic is about form, not meaning. Put differently, in logic, the meaning is in the form. Logic deals with general forms of reasoning, such as:

$A > B$
A
$\therefore B$

where the sign ">" stands for IF_THEN, and the three dots sign stands for "therefore."

The above syllogism is a quite general form of reasoning, in which the signs A and B stand for propositions, and the reasoning is independent of the meanings of the propositions. This form of syllogism is called *modus ponens*. The form is stripped of meaning, so the only relevant thing is the structure or the form, the internal connections between its components, and nothing else. As a cognitive activity, mathematical reasoning requires oblivion no less than memory. We must forget meaning and focus our attention on the abstract form. As a child, you may have been told to exercise repeatedly to remember certain mathematical facts. I guess that you have never been instructed to forget meaning in order to see the form underlying the particulars. Here we encounter again the idea of similarity. Regardless of different particular manifestations, *modus ponens* refers to the same general form of reasoning.

Let's stay with syllogisms for a moment. Think back to Boole and the idea of operations. Kauffman [3, p. 35] suggests that the entailment relation IF a THEN b can be expressed by Spencer-Brown's logic of distinctions. He explains that the entailment $a \to b$ equals $(\neg a) \lor b$. This means that entailment is equivalent to either a is not true or b is true. Therefore, if a is true, then b. As marking a (i.e., (a)) results in its negation $\neg a$, we have: $a \to b = (\neg a) \lor b = (a)b$. To get a better grasp of this idea, consider True (T) as the marked state $T = ()$ and False as the unmarked state $F = $. If a is True, then $\neg a$ transforms it to the unmarked state. The unmarked state, followed by the marked state b, results in a marked state.

Grounding syllogism in the logic of distinction is a wise move in expressing the similarity of differences. However, I use it only to show how similarity of forms is created and that mathematics involves identifying general forms that are similarities abstracted from particularities, whether the particularities of meaning or the particularities of visual appearance.

When Whitehead describes mathematics as the necessary inference of propositions from propositions, he actually points to a *structure* or *form* of reasoning where differences in the truth values of the propositions are ignored in favor of similarity. Our minds are tuned to identify structures as abstractions with significant informative value. A *structure*, or a pattern, is an abstract *relational* form connecting different elements. In Chap. 5, I will show you how to identify hidden meaning in a poem by drawing on relational structures. My emphasis on relations aims to orient the reader toward a different perspective on the mind and mathematics. This point will be explained and illustrated along the way.

The repetition of an ornamental element in an arabesque is informative as it orients us toward what we should expect. A violation of a pattern or its variation introduces newness and surprise. To conclude, the similarity of differences is a gateway for exposing deep patterns, forms, and structures underlying the activity of the mind in its different expressions. In the next section, I discuss one of these general patterns: repetition. I argue that repetition, a repeating theme in fields ranging from mathematics to poetry, is a structure indicating similarity. Understanding this point may help us to resolve several Kushiyot, such as how identity is formed and why, in his famous poem, Dylan Thomas urged his dying father to rage against death.

2.2 Play X Again, Sam

This section examines the idea of *repetition*, which we met in the previous chapter. To recall, Boole argued that repetition leaves our object unchanged. He expressed this idea as xx = x or x^2 = x. First, let us try to understand what repetition is. Repetition is the recurrence of an object, action, or event. Repetition lives in time and somehow describes a similar object, action, or event recurring along the time-line. This recurrence appears in different expressions of thought and human creation, from mathematics to poetry and architecture [4]. In mathematics, repetition has some simple expressions. For example, the most basic form of numbers, the natural numbers N = { 1, 2, 3, … }, arises from the act of *counting*. Counting itself is a repetitive process. Starting from 1, we repeatedly add 1 to obtain the next number. By adding 1 to the number, we simply and linearly increase the size of our collection and, correspondingly, the number that represents it. In this case, the repetition is the repetition of an operation. We repeat the *same* operation again and again.

It is fascinating to realize how repetition appears in different realms of human behavior. For example, in *poetry*. In Dylan Thomas's beautiful poem "Do not go gentle into that good night," the same line appears repeatedly, powerfully emphasizing the poet's feelings when experiencing the death of his father. Later, I will devote a section to repetition in poetry and show how we may gain a new interpretation of this poem through the idea of repetition as similarity.

We might conceive of repetition in a narrow way as the reproduction of a certain unit, as in the example of Fig. 2.1, where a pattern emerges from the repetition of the same basic object.

Repetition also appears in the sequence:

xxxxxxxxxx

and in the following sequence:

100100100100100100

Fig. 2.1 Pattern as repetition. (Source: Author)

Observing the sequence of x's, we can understand Boole's first law of thought and Spencer-Brown's law of calling. Repeating the same sign is a redundant way of saying the same thing: x. In fact, in terms of *information compression*, the long sequence can be compressed into the same sign repeated N times:

$$\text{xxxxxxxxxx} \rightarrow \text{x}$$

Repetition is not limited to the simple repetition of the same mark. Repetition can be the recurrence of a *function* (or several functions), such as in the case of an *iterated function*, where we repeatedly apply the same function. To recall Bohm [5], interesting things happen when we examine them from the perspective of dynamics and process rather than as given and complete objects. This perspective should not be confused with a philosophy of bleary-eyed philosophers trying to avoid the scientific approach. Bohm's perspective is deeply grounded in science. Take, for example, the foundational scientific concept of *energy*. Hewitt explains that [6, p. 151, my emphasis]: "We sense energy in things only when it is being *transferred or being transformed*." Shifting our attention from energy as an object/noun to energy as a verb/process is essential for understanding the natural world. We will get to this point later. In the meantime, consider how interesting it is to think about repetition as a dynamic pattern of the human mind.

We may find it interesting to think about repetition as an iterated function. Formally, if X is a set and $f: X \rightarrow X$, then f^n is the n-th iteration of f. Therefore, f^0 is the identity function on X, as nothing has changed yet, and f^{n+1} is $f \circ f^n$, which is function composition. Generally speaking, one starts with an initial value x_0 and applies the function n times. For example, consider the simple function $f(x) = 2x$. We apply the process described above, and the first iteration is:

$$(x_0) = 2x_0$$

The second iteration is:

$$(f(x_0)) = f(2x_0) = 2(2x_0) = 4x$$

And the third iteration is:

$$(f(f(x\,0))) = f(4x_0) = 2(4x_0) = 8x_0$$

For example, if $x_0 = 1$, then:
First iteration: $f(1) = 2$
Second iteration: $f(2) = 4$
Third iteration: $f(4) = 8$

and the resulting sequence begins 1, 2, 4, 8. Repetition as an iterated function brings us back to the extension of the first distinction. Through the inherited self-referential activity of the first distinction, our world can be expanded to include varied and complex forms.

One interesting aspect of an iterated function is that, by applying it, we may identify an *invariant set* that remains unchanged under the transformations of our set. Let me explain and illustrate how repetition may help us identify an invariant set. Why is this important? Boole shows us how identity is grounded in repetition, but here, I would like to show how an object's identity and the equality of objects can be constructed through repetition.

First, I have chosen *Brouwer's fixed point theorem* [7, p. 121] to explain and illustrate the idea of a fixed point. The theorem is presented as follows: "Let D be a closed disk … and f a continuous endomap[1] of D. Then f has a fixed point." The theorem states that any continuous function from a closed disk to itself has at least one fixed point. More formally, the theorem states that for any continuous function f mapping a closed disk D in R^2 to itself, there is a point $x \in D$ such that $f(x) = x$.

This theorem is easy to illustrate if we think for a moment about old vinyl records. Each record is a disk centered at the coordinates (0,0). The record is placed on the gramophone and starts rotating. Imagine that the record is a disk with no hole at the center. Each point on the disk gets moved to a new position as the record spins. In fact, it rotates through a certain angle, which specifies its new location. So, which point remains unchanged? Which point remains the same under this transformation? Correct. It is the point located at the *center* of the record. The center remains unchanged despite the transformations.

Here is another example. If you have prepared a cup of hot cocoa, you have mixed the powder in the water (or milk) and stirred it with a spoon until it has dissolved into the liquid. If you look at the cup from above, the particles of the cocoa drink may be imagined as particles moving on a disk. The fixed-point theorem states that no matter how much you stir the cocoa drink, at least one tiny particle ends up in the same spot it occupied before you began stirring. Even though the drink is well mixed, one small particle in the liquid will necessarily be found at exactly the same position. The fixed point can thus be described metaphorically as the eye of the storm. It is a place where things remain unchanged despite the changes the system experiences.

From the above examples, we learn that it is possible to define the *center* of a dynamical system as its fixed point or invariant set. This idea has far-reaching consequences. Think about personal identity and what makes you feel consistent and stable through the years. Your "self" is your center, the invariant that remains the same despite the transformations you have undergone throughout the years. You signify it while using the first-person pronoun "I." How can such a center be maintained despite the significant transformations we have experienced through the years? This is a fascinating question that I've studied in my previous work [8]. As a thought experiment, you may ask yourself under which transformations your self remains as a fixed point or an invariant set. From a broader perspective, we should

---

[1] A map from the set to itself.

realize that the existence of a center is one of the basic principles of mind and nature, as proposed by Alexander in his seminal work [4]. Here, we associate the idea of a center with the ideas of a fixed point, invariant set, repetition, and iterated function.

Here is another example used to illustrate the relevance of repetition for finding the invariant set or a fixed point. Let us consider a disk and a contraction mapping that shrinks the differences between points. We use the function $f(x) = \dfrac{x}{2}$ and apply it repeatedly. Let D be a disk with a radius of 1. The iterated function system involves repeatedly applying the function and mapping each point in the disk to a new position on the disk. Our starting point is $x_0 = (1,0)$, which sits on the disk boundary. We start applying the function as follows:

First iteration: $x_1 = f(x_0) = \frac{1}{2}(1,0) = (0.5, 0$
Second iteration: $x_2 = \frac{1}{2}(0.5,0) = (0.25,0)$

and so on. You can see that the process converges to the point located at the center of the disk. This is the fixed point. Nice, you may say, but what does all this mathematics have to do with human affairs? I now turn to dictionaries to show the relevance of repetition in the context of human artifacts.

2.3 What Is the Kernel of Our Dictionary?

Up to now, I have illustrated how repetition, as expressed by an iterated function, may help us identify the fixed point of a dynamic system. My interest is not in these instances per se but more natural expressions of the human mind. As the mind is mediated through signs, it is interesting to examine the role of repetition as a foundational mental structure in the realm of signs. For example, if you examine the dictionary, you will see that each word is defined through other words. This *hermeneutic circularity* may look like a labyrinth with no way out: a word is defined by other words, which are defined by other words, in a circular process that never ends. "Why is it a problem?" one may ask. The answer is that it is not necessarily a problem, just a matter of fact. This is the way the dictionary is constructed. The dictionary is an artifact, a product of human beings. As such, it is what it is. However, if the dictionary is supposed to represent our mental lexicon somehow, then defining one sign in terms of the others is a process forming a closed set, and we don't understand how a closed set of signs can transcend its closure and lead to something other than itself.

In this context, it is interesting to ask whether the collection of signs, as defined by the dictionary, has some stability, or an invariant set that functions as the system's core or kernel, and whether this kernel can somehow help us to understand how a closed set of signs can be used to represent something other than itself. In other words, identifying this core may help us understand how the set of signs can be used to mark things outside the set. This question has already been examined by [9]. Still, here, I use a different approach, showing how a repetitive iteration process may help us to resolve the circularity and mystery of our sign system. The general procedure is as follows:

Let S be a set of signs.
Let $D(s)$ be the definition of a sign $s \in S$.
Start with an initial sign s_0 (e.g., "cat").
Let $F(s)$ map the sign s to the set of signs used in its definition.
Repeat the process until a certain criterion is fulfilled.

Take the word "cat," for example. A cat may be defined as "a small domesticated carnivorous mammal with soft fur, a short snout, and retractive claws." We first identify the signs used to define the sign "cat":

$F(s_0)$ = {"small," "domesticated," "carnivorous," "mammal," "fur," "snout," "claws"}

Now, we iterate. For each sign $s \in F(s_0)$, apply F again to get $F(s)$, and continue this process iteratively. For example, the signs: small, domesticated, carnivorous, mammal, fur, snout, and claws all enter our machine that seeks their definitions and outputs the following:

"Small": Not large.
"Domesticated": Tamed and kept by humans.
"Carnivorous": Eating meat.
"Mammal": Warm-blooded animal with hair or fur, and females produce milk for their young.
"Fur": Soft, dense hair covering.
"Snout": The projecting nose and mouth of an animal.
"Claws": Sharp, curved nails on the feet of an animal.

We repeat the same process until we reach an agreed criterion such that the set is a set of self-explanatory or understood signs without further definitions. The output of this process may be the following signs: animal, hair/fur, meat, tamed, human, nose, and nails. Do you notice something interesting about this set? You can see that, with respect to this specific example, the kernel involves three body parts: hair, nose, and nails. It also involves two general categories of human and "nonhuman" (i.e., animal). This is not an exhaustive example, but the idea that our sign system is grounded in a small set of "embodied" signs seems well-founded and illustrated even through my specific example. In addition, it includes an important distinction between our kind (i.e., humans) and others (i.e., animals). The core of our mental lexicon is embodied and relevant to our most basic experiences as human beings. The invariant set of language is not the same as a mathematically defined invariant set. However, through repetition, we have reached an interesting core that may ground our sign system in reality, regardless of the particularities, dynamics, and noise in which this core is embedded.

2.4 Repetition, Identity, and Mathematics

What lessons can we draw from the above discussion? In the previous chapter, I explained that the equality sign "=" was first introduced by the Welsh mathematician Robert Recorde, who used it as a sign abbreviating the longer expression "is equal to." According to this historical process, human beings may have noticed similarities between things, and this similarity can work up to identity, as expressed by the equality sign. However, the basic and fundamental root of the mind is a distinction (i.e., a difference), a binary and qualitative difference. How can similarity be inferred in a realm based exclusively on differences? A cat and a cow are different creatures. When pointing to their similarity, through the fact that both have four legs, we are observing a higher-order activity of the mind where similarities and differences are noticed. In fact, identifying similarity means finding a *common denominator* between different objects. A child struggling to find the common denominator of two fractions and a Scotland Yard detective struggling to identify the common denominator of several crimes are engaged in identifying similarities and differences. This activity is far from trivial, as one discovers from attempts to create measures of distance or metrics. The many possible measures of distance are not necessarily related to each other. There is no simple measure of distance, and therefore, no simple measure of similarity. Similarity is not a trivial thing to identify, and neither is identity. Here, I would like to put aside these difficulties and focus on the equality sign.

The mathematical equality sign was invented as a name and abbreviation for the naïve use of language. The expression "is equal to" is a naïve, rather than a formal term, signifying similarity up to identity. Therefore, the sign "=" is a *substitute* for a naïve understanding and expression through language. It is a mathematical sign corresponding to an expression involving signs in natural language.

The equality sign marks a foundational axiom of human thought and mathematics. This is why I devote so much attention to understanding the mind-grounded meaning of this axiom. Describing nondifference using natural language, the expression is "x is equal to y." The etymology of "equal"[2] shows us that it emerged from *aequus* "on a level with." Two things that exist on the same level are equal. The Latin *aequalis* was also used in the sense of "uniform." To understand the deep association between equality and uniformity, have a look at the distribution of points in Fig. 2.2.

Is it a uniform or a nonuniform distribution? The units are the same except for their location on the board. They are different by not being the same. However, no spot can be differentiated from the others. It is as if the same thing has been produced through a simple algorithm of reproduction and repetition. What about their arrangement? We can describe the arrangement using arrows between the points, as shown in Fig. 2.3.

---

[2] https://www.etymonline.com/word/equal

Fig. 2.2 A uniform
distribution of points.
(Source: Author)

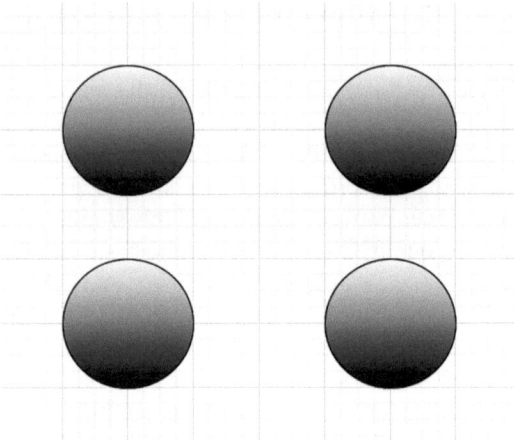

Fig. 2.3 The same spots
but with interconnections.
(Source: Author)

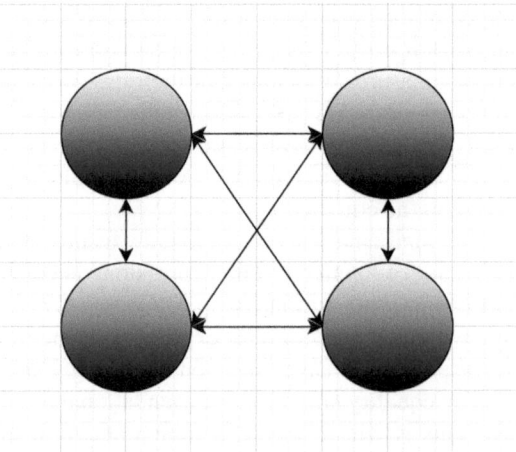

We can see that each spot (i.e., unit) has the same number of arrows entering and leaving the spot. We cannot distinguish between the units. They are the same units, and their location within the whole does not help us either. In fact, we observe in the figure a perfectly symmetric structure, whose identity is preserved under well-defined *transformations* of the spots. An interesting association has now been formed between transformation, symmetry, and identity.

At the most basic level, equality and identity are grounded in nondifference and *symmetry*. These ideas can be understood from the perspective of information theory. The *graph entropy* of the above figure is maximal because there is no way we can distinguish between the spots. Identity is, therefore, expressed in nondifference, uniformity, symmetry, maximal entropy, and uncertainty, all grounded in a simple computational process of repetition. When one cannot see the difference between this and that, they are considered the same or similar to a certain degree.

2.5   Naming and Imagination

What happens if we *name* each spot? Take a look at Fig. 2.4.

Marking each unit with a sign forms a difference and decreases the information entropy of the graph to a minimum. Although the units are indistinguishable and their relations and arrangement are the same, we can now distinguish between them. Naming each spot with a different letter has now made them different. Our sign system can, therefore, generate differences that are not grounded in the system we observe: *Differences that are not grounded in reality*. This is a highly important point as it explains mathematical imagination. In the next chapter, I will mention A.N. Whitehead and the importance he gave to *imagination*, which he considered in terms of *realizing possibilities beyond the given reality*. When using signs, we can extend the realm of possibilities, such as the possibility of a new differentiation that exists beyond the given reality. The above figure shows that the spots cannot be visually differentiated. This means that, in reality, they are the same. By giving them names, I created a difference not grounded in reality.

I find this idea quite remarkable: *imagination explores possibilities through the mediation of signs*. It is an idea that occurs repeatedly in the insights of great authors. For example, Cormac McCarthy wrote in one of his latest novels [10, p. 15, my emphasis]: "Nobody comes with names. You give them names so that you can *find them in the dark*." This insight corresponds perfectly to the discussion so far. Signs, whether in natural language or in mathematics, are not just a way of naming collections of objects. Signs are also marks for identifying things "in the dark." They help us imagine possibilities beyond our shortsighted vision and its grounding in the sensorimotor experience.

Here is an example illustrating signs and imagination. Do you remember the idea of the *imaginary number*? An imaginary number includes an *imaginary unit* i: $i = \sqrt{-1}$. It is the solution to the equation: $x^2 + 1 = 0$. So, the unit i is equal to the square

Fig. 2.4 Named spots.
(Source: Author)

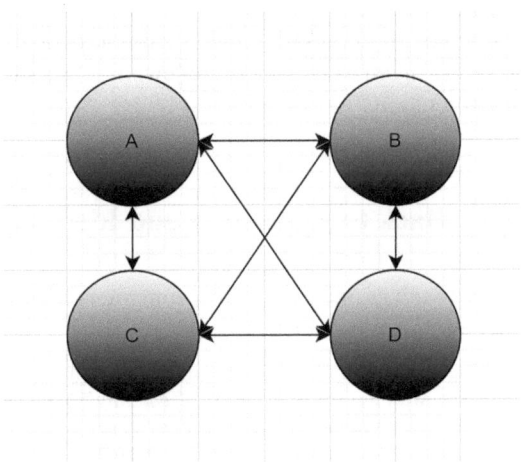

root of −1. But wait a moment! No real number squared will result in a negative number! So, what kind of creature is *it? Mathematical creatures, the same as other man-made* creatures, are formed by the operation of the human mind. The sign *i* is not a mark of a *perceptually* marked distinction. No one ever observed *imaginary units* hanging from trees or wandering through the meadows. It is not a distinction existing in the same way as my cup of coffee or the computer I am using to write this book. The imaginary unit *i* is just one example of *a sign forming a distinction in the realm of mind possibilities.*

This idea is difficult to accept as it suggests that our mind forms objects rather than reflects them. I first learned about this counterintuitive idea when I was an undergraduate student. A short book by my professor of psychophysics explained that we usually think of the measurable object as preceding the act of measurement. For example, as young children, we learn to count using our fingers; we form a one-to-one correspondence between the units of measurement (i.e., the numbers) and the measured (counted) objects (i.e., the fingers). The counted fingers precede the act of measurement and are independent of it. You don't have to count your fingers to realize their existence. Your fingers exist whether you count them or not. However, the idea that the measurable object precedes the act of measurement is expressed only in the first phase of human and scientific development, whether on the level of evolution, personal development, or cultural development. My professor explained that the history of science proves that our measurement operations define the measured object and precede it.

How much freedom do we have to define an object through our operations? Here, we get into the pragmatic aspect of science. Our imagination presents possibilities beyond the given, but from a scientific perspective, its products are judged when we return back to earth. For example, kinetic energy is an outcome of human imagination. Still, since we can define it mathematically, measure it, and use it, this suggests that something exists that corresponds with our imagination. How do we know? When the idea of kinetic energy is put back to work, we can use it to explain, predict, and control events in the world. As argued by Whitehead [2, p. 7]: "The true method of discovery is like the flight of an aeroplane. It starts from the ground of particular observation; it makes a flight in the thin air of imaginative generalization; and it again lands for renewed observation rendered acute by rational interpretation."

2.6 Back to Repetition and Identity

As individuals and collectives, constructs of our mind are far from trivial. For example, as argued by Mazur [11, p. 1]:

> One can't do mathematics for more than ten minutes without grappling, in some way or other, with the slippery notion of *equality*." Why "slippery"? Because "the way in which objects are presented to us hardly ever, perhaps never, immediately tells us—without further commentary—when two of them are to be considered equal (ibid.).

At this point, we can better understand the equality sign as a far from trivial *process*, far removed from basic intuition. Indeed, it has been shown that children [12] have great difficulty understanding the meaning of equality. For instance, first graders conceive it as "what it adds up to." This naïve knowledge explains why children may struggle to solve equations such as $1 + x = 6$. If the sign "=" points to what "adds up" and a part of the components forming the addition is in question (i.e., the sign "x" or an empty space "_"), then it is not clear what adds up.

Identity, or in broader terms *equivalence*, is not trivially given to the human mind beyond what can be gained by a simple process of sensorimotor comparison. It is a sign indicating the equality of value, and value, as I already explained, is not in the sign but in the *mapping process*. Therefore, when presented with the expression $3^2 = 9$, the school student should understand that the value is the same despite the different appearances of the marks on the right and left sides. It is the same in the sense that raising 3 to the power of two results in 9, being the same number as the one to the right of the equality sign. *Equality is a sign pointing to a symmetry under certain transformations.* This is probably the main lesson I want you to take from this chapter. It shows how identity is formed in different expressions of the human mind and cultural products, from mathematics to poetry.

With respect to $3^2 = 9$, we can see that the symmetry of the right and left sides of the equality mark must somehow be identified to justify its meaning. In contrast to signs that indicate a class with a sensorimotor grounding, such as the sign "cat," the equality sign has no simple sensorimotor class of references. It is grounded in various experiences where things appear on the same level, where they appear to be uniform, where they look the same, and so on. It marks a nontrivial value that cannot be grounded in a single, basic sensorimotor experience.

A possible hypothesis that I would like to present is that one function of repetition is to *fix the identity of an object by exposing transformations that do not change the object*. To present the hypothesis, let us return it to Boole and the importance of 1 and 0. Today, Boole's binary numbers would probably have been excluded by WOKE culture. "Reality is nonbinary," some snowflakes from an Ivy League university would have screamed, "and your numbers are nothing more than the representation of an oppressive binary hegemony imposed by a capitalist colonialist regime led by privileged white men." Some "elite" university presidents might even have ordered an emergency meeting to discuss excluding Boole's book from the library. However, here, we will keep our minds open enough to include the possibility of binary objects whose existence is deduced from the most basic form of the human mind: the primary distinction.

Let us prove that 1 raised to the power of 2 equals 1 using an iterated function. We consider the function $f(x) = x^2$. Let's apply it for $x = 1$.

$f(1) = 1$

$f(f(1)) = f(1) = 1$

$f(f(f(1))) = f(f(1)) = f(1) = 1$

and so on.

This demonstrates that 1 is a fixed point of the function $f(x) = x^2$, *stabilizing* the identity of x when it is equal to 1. The fixed point, or the invariant set, stabilizes the identity of our objects, i.e., their equality value. What does this mean? It means that 1 is the number that remains the same when we repeatedly use Boole's $x = x^2$. The kernel of our number system includes the number 1, which is, in its most general sense, equivalent to the terminal object or the Universe.

Returning to Boole, we are now in a better position to understand the importance of 1 and 0 when aggregating the same sign by repetition through the operation $xx = x$ or $x^2 = x$. The identity of x is "stabilized" through the fixed points 1 and 0. The magic sauce in understanding the identity sign lies in a process or a transformation which, when applied to the object, enacts its identity for a certain observing system. This process is grounded in a binary universe of mutually exclusive distinctions: 0 and 1.

At this point, we may also better understand the developmental difficulties involved in understanding identity and equality. We think from bodies, and the developmental perspective proposed in [13] suggests that we should be sensitive to the development phase. One thing that clearly characterizes the development of human beings is that they change very quickly when they are young. The details are less important than the observation and its consequences. To understand the mathematical notion of identity, the observer cannot itself be in a chaotic flux of change. How can a child understand identity when he, as an observer, is in a process of transformation? Can the caterpillar define its identity when transforming into a butterfly? Identity is not formed outside the observing system but through the observing system, as proposed by von Foerster's idea of *eigenvalues* [14]. The objects we observe, argued von Foerster [14], are "tokens for stable behaviors." Objects we believe to exist independently of our minds are, in fact, precisely examples of stable macro-level behavior observed on the surface of underlying and unfolding dynamics. Indeed, the above is nothing but a repetition of von Foerster's argument. A stable piece of behavior is not easy to identify when the objects are observed through an observer-in-transformation. This is probably one reason why mathematics cannot be taught to 3-year-old infants. The idea of repetition as a process stabilizing identity also finds its expression in the world of good old Sigmund Freud, as explained in the next section.

2.7   Freud: "Remembering, Repeating and Working-Through"

One cannot discuss the idea of repetition without mentioning Freud's classic essay: "Remembering, Repeating and Working-Through" [15]. Freud, the sensitive observer, noticed that sometimes a patient with childhood trauma does not remember anything of "what he has forgotten or repressed" [ibid., p. 150] but "acts it out." The patient does not reproduce the memory but reproduces the painful event as "an action." In Freud's terms, "he *repeats* it." Repetition, so argued, *replaces remembering* and functions as a mechanism for *unearthing repressed memories*. Surprisingly, Freud's general thesis corresponds to the mathematical approach presented before, and this correspondence can be explained through Post-Traumatic Stress Disorder (PTSD).

PTSD occurs in some people who have experienced a painful or scary event. One observation is that the difficulty in "digesting" the emotional intensity of the event results in an out-of-body experience, where the person experiences the event as if it happened to someone else. The difficulty in digesting the event results in unprocessed memories. These memories may be described metaphorically as food not ·digested by the body. When painful memories exist in their raw form, they cannot be tolerated, owing to their emotional intensity. They cannot be appropriately represented or named in a way that allows the person to tolerate them. Think about the horror of encountering an entity that cannot be named. This horror appears repeatedly in our cultural artifacts. In Harry Potter, "He-Who-Must-Not-Be-Named" or "You-Know-Who" is Lord Voldemort. The ultimate darkness of this character is something that is beyond the sign-representation of the other characters. The experiences of a patient with PTSD do not exist as representations that can be digested. They are raw realities that keep torturing the poor individual. The painful memories are like the ghost of a dead man not properly buried. They never stop haunting the patient.

In this context, as a mechanism for unearthing the memories, repetition may be interpreted as a desperate attempt to produce a "token of stable behavior," where my emphasis is on the idea of a "token" or a representation stabilizing the identity of the thing that is beyond our grasp. It is an attempt to ground the raw representation of the event in order to stabilize it and digest it. This hypothesis, which is based on the mathematics of repetition and its importance in identifying the "fixed point" of the system, is that repetition as a process aims to affirm the identity of an event that cannot otherwise be stabilized as an object. This point may be difficult for those with a simple representational theory of the mind. However, a novelist's insight may help them understand this point better.

In *The Book of Illusions* [16], Paul Auster describes the transformation from the old silent movies to modern movies with their hyper-real representation. He insightfully observes that [ibid., p. 12, my emphasis] "… the paradox was that the closer movies came to simulating reality, *the worse they failed at representing the world—* which is in us as much as it is around us." This is an insightful observation. It tells us that *the more realistic the representation is, the less it represents reality*! This observation sounds like a contradiction. Another Kushiya for Monsieur Chouchani. How is it possible that the more realistic the representation is, the less it represents reality? Isn't it supposed to be the other way around?

The explanation is that the represented world is "in us as much as it is around us." For us, *reality is the reality of representations*, not the one they represent, as if reality exists independently of our minds. This is the reason why patients with PTSD suffer so much. In a simple representational system, the traumatic event would have been registered in their minds like an item represented in the huge inventory of

Amazon warehouses. But this is not the case. There is no registration without a representation according to the world within us, and most of us, as surprising as it may sound, cannot acknowledge the reality of real and overwhelmingly painful events or nondigested reality in general.

This idea emphasizes the important connection between representation and reality. The idea can be illustrated through another insightful observation by the author Steven King. In his novel *Misery* [17], the hero is an author who suddenly realizes that [ibid., p. 307]: "Truth isn't stranger than fiction, no matter what they say. Most times, you know *exactly* how things are going to end." Why do you know how things are going to end? And why are truth or reality no stranger than fiction? Literature represents reality, like any other form of semiotic representation, and is distant enough from reality to turn it into a familiar realm. Truth is not stranger than fiction because it is only through fiction (i.e., the activity of sign systems) that we can approach reality and find things in the dark. The same logic of the mind is evident whether we observe literature or mathematics.

2.8   Repetition in Poetry

As promised, I will discuss the abstract form of repetition as expressed in poetry. Repetition is a common device used by poets to address different aims, from emphasizing a theme to creating a rhythm. However, I would like to point out that repetition in poetry is also associated with similarities and equivalence. We may gain new interpretations and understandings of a poem by considering this possibility. To explain and illustrate this point, let's read the first lines of Dylan Thomas's famous poem "Do not go gentle into that good night":

Do not go gentle into that good night,
Old age should burn and rave at close of day;
Rage, rage against the dying of the light.

In line 1, "Do not go gentle into that good night," Thomas urges an unspecified addressee (probably his dying father) not to accept death passively or peacefully. The night is a known metaphor for death, and while the night is described as "good," the night in the poem is still a metaphor for death. Here, the first Kushiya pops up. The poet asks his father NOT to accept death. But if death/night is "good," what is the problem with slipping gently around to the dark side of the moon? Let's try to resolve this Kushiya by analyzing the similarities forming repetitions.

We can see that "night" rhymes with "light." The rhyme, therefore, establishes a link between two different words. This link establishes a possibly unnoticed and unconscious *similarity* between the two different words: "night" and "light." When night and light are conceived as similar, we observe a hidden repetition of the same/similar unit despite their differences. This is an unfolding process. Two different words are connected by a rhyme. Once connected, they are considered the same.

The recurrence of the same unit in the poem invites us to examine the similarity and to use it for inference.

While night represents death, light represents life. Therefore, we do not observe a simple repetition of the same object but a repetition accompanied by variation. Night and light are similar through rhyme, but different as oppositions. The similarity forms a repetition, but the repetition also points to the two connected words as oppositions. They are two faces of existence, the opposition of life and death. However, by unconsciously establishing the similarity between night and light, death and life, we understand that the "good night," meaning "good death," is also "good life." The same theme appears in line 4 of the poem: "Though wise men at their end know dark is right." "Dark" is a synonym for "night" and "death," and "right" rhymes with "light." The synonym is another point of similarity, establishing a repetition. Therefore, a similarity is established between dark, night, and death, and between light, life, and right. By inference, if dark is right, then dark is light, and death is life. *Inference through substitution* of similar objects is an interesting reasoning process characterizing our mental realm as children and poets.

Repetition, as previously explained, can be used to stabilize an object's identity. Here, it points to the similarity of opposition and connects different units to form a hidden structure.

At this point, we may use similarity to resolve some problems. For instance, why rage against the dying of the light? The answer is allegedly given in line 5: "Because their words had forked no lightning they." A possible interpretation is that wise men might feel that their words or deeds did not make a significant impact ("forked no lightning"), which might compel them to resist death. They (line 11) "… learn, too late, they grieved it on its way," meaning they regretted the passing of time ("grieved it on its way") instead of fully appreciating it. This interpretation does not resolve the Kushiya. Let us try a different direction guided by the similarities and repetitions.

The poet addresses his father, asking him (Line 17): "Curse, bless me now with your fierce tears, I pray." The poet seems to ask his father for two opposing but conflating ideas: bless and curse. The call to rage against death, despite the fact that it is described as "good," may now be better understood. Life and death are two aspects of life, like darkness and light. Through the similarities/repetitions, we learn that they are the two complementary facets of the same distinction. However, an underlying meaning is exposed when we remember that the poet is in a tantalizing emotional state, where he asks the father to curse/bless him with his fierce tears. The son, urging his father to struggle against death, is praying for his father's blessing and is actually using the old sense of blessing from the Old English *blet sunga*, which means blessing, as "divine grace; protecting influence." The etymology of "bless"[3] teaches us that it has originated from "speak well of." The father, passing from one side of existence into the other, is asked to fulfill his archetypal role in protecting his child [8]. Then, why curse?

---

[3] https://www.etymonline.com/word/bless

When Dylan Thomas uses the word "curse," which is the opposite of bless, and prays for his father to bless/curse him, he is asking for his father's protective influence on the dark side of existence and expressing the fear of the curse originating from late Old English in the sense of "a prayer that evil or harm befall one; consignment of a person to an evil fate." The Kushiya can be resolved by interpreting through similarity, repetition, and inference through substitution. The reading of the poem resolves the Kushiya by pointing to the poem as a unique expression of a general human theme: the son asking for his father's blessing while being afraid of his curse, specifically when the father is transitioning to the darkness of death. In other words, the similarities between different entities in the poem are repetitions that help us identify a pattern. However, to understand why the poet is so desperate to ask his father to rage against death, we need to understand the opposition between a curse and a blessing that appears at the end of the poem. A son asking for his father's blessing is a repeating theme in human culture, and the complementary theme is the fear of being cursed by the father's power. Human thought is grounded in basic distinctions forming complementary oppositions. Associations are made between seemingly unrelated objects through similarities and repetitions, combining several objects into the same set or collection. The discovery of the same dynamics in mathematics and unrelated disciplines, such as poetry, leads to an interesting nexus.

Aharoni [18, p. 6] suggests that "… a sense of beauty in poetry and mathematics is the result of our immediate and unconscious response to a hidden structure." The structure of identity is exposed through repetition, and the structure of Thomas's poem is at first hidden but exposed through repetition. The son, faced with his father's death, asks him to rebel against the passage to the dark side of reality. This passage is accompanied by danger. The dead can curse the living, and the son asks for his father's blessing before the passage is completed. A repeating theme in human culture is expressed in the poem through a form of repetition that points to deep similarities. Here, another nexus is established between mathematics and the mind through underlying structures, forms, and dynamics.

Points to Note
- **Repetition**: points to similarity.
- **Repetition as an iterated function**: helps us to understand fixed points and invariant sets.
- **The center of a dynamic system**: can be defined by fixed points or invariant sets.
- **Boole's law of thought**: is an expression of a repetition.
- **Transformation and symmetry:** are deeply associated with identity.
- **Realistic representation**: is paradoxically less representative of reality.
- **Inference through substitution**: is an important mental device in mathematics and other expressions of the mind.

References

1. Verhulst, F.: Mathematics is the art of giving the same name to different things: an interview with Henri Poincaré. Nieuw Archief voor Wiskunde. Serie 5. **13**(3), 154–158 (2012)
2. Whitehead, A.N.: A Treatise on Universal Algebra: with Applications. Cambridge University Press, Cambridge (1898)
3. Kauffman, L.H.: Laws of form, a survey of ideas. In: Kauffman, L.H., et al. (eds.) Laws of Form—A Fiftieth Anniversary, pp. 1–88. World Scientific Pub Co, Singapore (2023)
4. Alexander, C.: The Nature of Order. Center for Environmental Structure, CA, Berkeley (1980)
5. Bohm, D.: On Creativity. Routledge, London (2004)
6. Hewitt, P.G.: Conceptual Physics, Thirteenth edn. Pearson, UK (2023)
7. Lawvere, F.W., Schanuel, S.H.: Conceptual Mathematics: A First Introduction to Categories. Cambridge University Press (1997)
8. Neuman, Y.: Conceptual Mathematics and Literature. Brill, Leiden (2020)
9. Massé, A.B., et al.: How is meaning grounded in dictionary definitions? arXiv preprint arXiv:0806.3710. (2008)
10. McCarthy, C.: Stella Maris. Alfred A. Knopf, New York (2022)
11. Mazur, B.: When is one thing equal to some other thing? In: Gold, B., Simons, R.A. (eds.) Proof and Other Dilemmas, MAA Spectrum, pp. 221–241. Mathematical Association of America (2008)
12. Rittle-Johnson, B., Matthews, P.G., Taylor, R.S., McEldoon, K.L.: Assessing knowledge of mathematical equivalence: a construct-modeling approach. J. Educ. Psychol. **103**(1), 85 (2011)
13. Luria, A.R., Vygotsky, L.S.: Ape, Primitive Man, and Child: Essays in the History of Behavior. CRC Press, Boca Raton (1992)
14. Von Foerster, H.: Understanding Understanding: Essays on Cybernetics and Cognition. Springer, New York (2007)
15. Freud, S.: Remembering, repeating and working-through (Further recommendations on the technique of psycho-analysis II). The Standard Edition of the Complete Psychological Works of Sigmund Freud. **XII**, 1911–1913 (1914)
16. Auster, P.: The Book of Illusions. Picador, New York (2002)
17. King, S.: Misery. Hodder and Stoughton, London (1987)
18. Aharoni, R.: Mathematics, poetry and beauty. J. Math. Arts. **8**(1–2), 5–12 (2014)

Chapter 3
Mind and Mathematics
in an Event-Centered Approach

Abstract The mind isn't reducible to sensorimotor experiences, and neither is mathematics. In this chapter, I argue that sign-based processes are the foundations of both mind and mathematics and explain this idea through the concept of the variable, which allows us to imagine and explore multiple possibilities beyond the concrete. The mind is presented through an event-centered approach, according to which *events, interactions, and transformations* should be the focus of our analysis. This thesis is illustrated by word meanings and multiplication. The chapter also discusses linear transformations, liquid cats, and the reasons why narrow-minded educators may be dissatisfied with genies.

3.1 On Irreducibility

In the previous chapters, I presented the idea of the primary distinction as the root of the mind. I also explained that concepts in our minds are formed by collections of things grouped together through similarity. Moreover, I explained that number is actually the name we give to a collection of collections presenting a similarity (i.e., isomorphism) regardless of their different appearances at the level of analysis of the collections themselves. I also discussed the notions of equality, identity, and repetition, and the way these general notions arise in different expressions of the human mind. By pointing out the expression of the same forms in various fields, I aimed to show how the mind is expressed in different domains and how mathematical forms are reflected in our mind. One should not confuse this approach with any attempt to reduce the various expressions of the human mind to mathematical concepts. That approach has been tested by great minds such as Piaget but ended in failure. My aim is to point to the expression of similar structures in both domains. However, the mind is not reducible to mathematics. To illustrate this point, let me use *set theory*.

In his classic Naïve Set Theory [1 , p. 1], Paul Halmos argues: "The mathematical concept of a set can be used as the foundation for all known mathematics." One should appreciate the significance of this statement, which says that the "concept of

Y. Neuman, *Mindmatics*, Mathematics in Mind,
https://doi.org/10.1007/978-3-031-74955-1_3

a set" is the foundation of mathematics. The simple idea of a collection of things is the foundation underlying the tower of mathematics with all of its branches. This foundational notion of a set is also expressed in the study of cognition, where cognition is presented as categorization. As argued by Harnad [2]: "To cognize is to categorize: cognition is categorization." A category is a mental set of objects. If a set is a foundational object of mathematics and a category (i.e., a mental set) is the foundational object of human cognition, then we may ask whether we can learn about the mind through set theory and vice versa.

To recall, we define a set through its members, and we need to understand what belongs to the set. However, set theory proposes that the concept of "belonging" is a "primitive" and therefore undefined [1, p. 2]. This is an important point. The idea of a set is a foundational concept in mathematics. However, this foundational concept is grounded in the *relation* of belonging, which is primitive, meaning primary and hence *undefined*. Therefore, categorization as the *raison d'être* of human cognition cannot be grounded in set theory, where the most basic notion of belonging is undefined and unexplained. The reason is that grounding categorization in set theory should serve some explanatory purpose. If the most basic relation of belonging is undefined, then there are no benefits to this reductive move.

Let me put it differently. If x is a member of set A, we describe the relation as *belonging* and say that x is *contained* in A. This relation can be signified as follows: $x \in A$. It is a "primitive" and "undefined" relation, which means that the foundations of mathematics are grounded in a relation whose nature is beyond the scope and interest of mathematics. There is nothing wrong with this approach of using an undefined primitive. In fact, this approach is inevitable unless one happens to enjoy being caught up in an infinite circle of definitions of definitions. It is not only an inevitable approach but one that has been shown to be highly constructive. As I explained in the first chapter, our interpretation must start somewhere, and extending the limit lines of our thought is possible only within some boundaries that never intersect with any kind of "totality" or "whole." However, to understand the mind, the question of "belonging" cannot be answered by assuming a primitive and undefined concept. It is a question requiring an answer in terms of foundational processes of the mind.

Here, the difference between studying mathematics and studying the mind becomes apparent. For set theory, the fundamental relation of belonging is an undefined primitive. For the study of the human mind, it is the object of study. A possible solution to the problem may be to reduce the relation of belonging to a more basic sensorimotor experience that can explain both mind and mathematics. If this is the case, then set theory can be reduced and explained through the mind. Let us examine this possibility.

3.2   In Embodiment We Trust

One may argue that the primitive notion of set membership can be embodied in basic sensorimotor experiences. According to this logic, *containment* is grounded in our primary experience of things within and outside us. When a child eats a piece of chocolate, the chocolate is digested and contained in him. When the child disposes of unnecessary materials from her body as feces, what was once inside and contained within the body is now on the outside. Can we reduce the set theory primitive of "belonging" to the embodied experience of "containment"? There is an explanatory power to the theory of the embodied mind, as advocated in [3], but there are also severe limitations. My cats are involved in containment activities every day. Still, to the best of my understanding, and despite my intensive efforts to teach them set theory for 15 years, they do not express the slightest understanding of set membership, at least as Halmos explains it. The point is that, despite the foundational activity of containment shared by humans, cats, and the nematode, one kind has developed set theory while the others have not. Explaining the mathematical primitive of belonging through the embodied experience of containment seems to be a limited theoretical move.

To refute my criticism, there are two possible counterarguments. First, one might say that we are not talking about *reducing* the meaning of a mathematical primitive to an embodied experience, but about *grounding* a mathematical primitive in a basic sensorimotor experience. This is a much better argument, as it points out that, developmentally, our embodied experiences somehow play a role in forming abstract concepts. This is an idea I accept and use. However, we face the following question: how do we bridge the gap between the basic sensorimotor experience and the abstract concept? The second possible counterargument aims to address this question.

This counterargument claims that the set theory relation of belonging is not grounded in containment per se, but in the *metaphorical use of containment*. According to this argument, the primitive of belonging is grounded in containment, but it is formed by transcending the basic embodied experience through metaphor: belonging is like containment. Here, we return to the dynamics of similarities and differences discussed previously. To move from the basic notion of containment to set theory, one must assume a *bootstrapping mechanism as a primitive mechanism of the mind*. It means that the real primitive underlying the abstract notion of set membership is not the sensorimotor experience of containment, but a rather nonprimitive and sophisticated mechanism of identifying similarities and differences. This idea contrasts *sharply* with the idea of the embodied mind, where the most basic units are sensorimotor experiences. In other words, although higher-level and more abstract concepts may be grounded in basic sensorimotor experiences, a thesis that I can easily see as productive, the basic mechanism that we need in order to understand how abstract concepts are formed is not embodiment but an abstract and sophisticated mental process we must assume in advance. My argument can be further explained through the mathematical notion of relation and its grounding in set theory.

Once we establish the notion of a set, its primitives, and axioms, we can turn to the notion of *order*. "What does it mean to arrange the elements of a set A in some order?" asks Halmos [1, p. 22]. Let us order the set {a, d, b, c} in alphabetical order {a, b, c, d}. Like any other set, the original set does not include any sense of order. It is only a bag of elements. Halmos uses a nice trick to explain how order can be described. He proposes that, for any "spot" in the ordering, we consider the set of all those elements that occur at or before that spot. Using this trick, we get the following sets:

$$\{a\}\{a,b\}\{a,b,c\}\{a,b,c,d\}$$

The collection that contains these sets as its elements is, therefore,

$$C = \{a\}\{a,b\}\{a,b,c\}\{a,b,c,d\}$$

"Which set is included in all?" asks Halmos. The answer is {a}. Therefore, it is the first element of the set. Now, we look for the next smallest element of C and ask the same question. The set identified is {a, b}, and therefore, it is second in our list, and so on. The reader may be struck by this idea of order, arguing that describing order in terms of set theory shows that containment is the basic sensorimotor experience underlying our most basic notion of order. However, as Halmos explains [1, p. 23, my emphasis]:

> [...] we may *not know* precisely what it means to order the elements of a set A, but with each order we can *associate* a set C of subsets of A in such a way that the given order can be *uniquely recaptured* from C.

Halmos should be praised for his precise language. He explains that order can be *represented* through the language of set theory, in the sense that we can *associate* each order with a certain set C so that the given order can be "uniquely recaptured from C." However, to decide what elements "occur at or before" a chosen spot, we must already possess a basic notion of order in the most primitive sense of before and after. That is, order cannot be reduced to containment, even when a similar idea of set inclusion is used in set theory to represent rather than to explain the idea of order.

As you can see, the foundational theory of mathematics cannot save us from foundational issues of the mind, and it has no pretensions of doing so. Mathematics cannot provide us with the foundations of the mind. The same holds true for the mind, which cannot provide us with foundations for set theory through any reducible move toward basic sensorimotor experiences. A different theoretical approach must be found.

3.3 Back to Signs

What underlies mind and mathematics, paradoxical as it may seem, are highly abstract processes involving *signification*. Naming a category of perception and formulating the axiom of specification [1, p. 6] are grounded in a highly sophisticated and abstract use of signs. There are different types of signs. Whitehead argued [4] that in "mathematical calculus," a sign is *substitutive*, meaning that it takes "the place of that for which it is substituted." There is a difference, Whitehead argued, between words in natural language and substitutive signs in mathematics. A word is an instrument for thinking about the meaning that the word expresses. In contrast, a substitutive sign is "a means of *not thinking* about the meaning it symbolizes" [ibid., p. 5, my emphasis]. As strange as it may seem, mathematical signs urge us to forget far more than to express the sign's meaning. One may find it challenging to understand this argument. When we use the sign of equality "=," isn't it to express the idea that x is equal to y? Whitehead's proposal shifts our attention to a processual way of thinking, where the meaning of signs is not determined by pointing to certain objects, but by substitution. For substitution, we need some rules to guide us in the process. When defining calculus, he writes:

> In order that reasoning may be conducted by means of substitutive signs, it is necessary that rules be given for the manipulation of signs. The rules should be such that the final state of the signs after a series of operations according to the rule denotes, when the signs are interpreted in terms of the things for which they are substituted, a proposition true for the things represented by the sign.

Whitehead's language is difficult to understand, but we may appreciate his argument better when he returns to Boole's definition of a sign. Mathematical signs, as arbitrary marks, gain their meaning not as expressive signs but as *signs of equivalence,* one may say. Their meaning is substitutive rather than expressive, and their substitutive character is gained through the operations and rules applied to them. Consider, for instance, the expression:

$$x + 1 = 2$$

What is the meaning of the sign "x"? The sign is a *variable*. Does it have a simple, expressive sense like the word "cat"? The variable is a sign that represents an unknown or changeable value. However, if the sign represents the unknown, how should we understand the meaning of this "unknown," which is by definition "unknown"? This Kushiya can be resolved by understanding that the meaning of x in the above expression is *substitutive*. By rule, we convert the expression:

$$x + 1 = 2$$

into

$$x = 2 - 1$$

to gain:

$$x = 1$$

Just as the proof of the pudding is in its eating, *the sign's meaning is in the transformative process*, resulting in the understanding that x can be replaced by 1, given certain rules or operations defined for the domain. Contrast the "meaning" of x in the above expression with the "meaning" of x in the expression:

$$x + 1 = 1$$

In this case, the "expressive" sense of x is different. It is not 1 but 0. Here, we understand that x does not stand for a thing, at least a priori. We can understand it only as a substitute for gaining meaning through a transformation process according to well-defined rules. The variable, therefore, represents a *potentiality* that turns into an *actuality* through manipulation according to well-defined rules. In Aristotle's philosophy, potentiality refers to the capacity to develop into a particular state, while actuality is fulfilling that potential. In the above expression, the variable represents potentiality, as it can take on any number of values within its *domain*, whereas a specific number or value represents actuality. But how do we know what the domain is? The domain is actually identified through rules and operations as part of a social convention. Such epistemic closure seems to reappear wherever we go, and here, a deep similarity appears once again between language and mathematics.

3.4   Variables in Mathematics and Language

The variable x introduces indeterminacy and ambiguity into our world. At first, it is an unknown thing that can be substituted for different things. It is like the ambiguity of words in natural language. For instance, the sign "I" is a variable that signifies nothing specific. It is not similar to the word "tree," which signifies the collection of trees, or to the sign "cat," which signifies the set of creatures ranging from my own cats to the imaginary character of Garfield the cat. The sign "I" seems to function as a variable in what may be metaphorically described as the "calculus of language." The meaning of "I," as the ultimate variable of natural language, is determined only by reference to the agent who uses it to refer to himself.

Variables are crucial for all aspects of mental activity from natural language to mathematics, as they allow us to imagine and explore multiple possibilities. In contrast, with signs naming a class, such as the signs naming natural numbers, variables are signs allowing much more freedom from a given reality. This is why algebra is a step ahead of arithmetic, as algebra provides us with more freedom to imagine different mathematical possibilities. In algebra, we have the idea of a variable as a name that does not specify a more concrete and particular reference, such as in the case of natural numbers.

Moreover, variables blur the dichotomy between static and dynamic. Variables can be seen as dynamic, capable of changing values, or static, representing a fixed but unknown value. This dual interpretation can lead to discussions about change, process, and stability in the nature of the mind and mathematical systems. At this point, our discussion naturally progresses toward realizing the importance of understanding the mind and mathematics in dynamic terms. For instance, the "meaning" of a variable is determined only on the fly. One cannot consider it to correspond to a stable and already-formed aggregate. We will discuss this critical point in the concluding section of this chapter.

Let us return to the sign "=." Whitehead explains that "two things are equivalent when, for some purpose, they can be used indifferently" [4]. Here, we return to the basic notion of a difference. However, Whitehead adds a qualification by saying that the equivalence of distinct things implies a definite purpose and a "certain limitation of thought or of action." Equality has no universal meaning, explains Whitehead. Equality or equivalence is meaningful only given the specific rules of the game.

This idea reminds us of Bateson and his hypothesis that the basic unit of the mind is a difference that makes a difference [5]. A difference existing in reality independent of the mind is a difference that is not accessible to the mind. A difference exists only if it makes a difference on another level of analysis [6]. When signifying, we express this idea. For example, the set of mammals is not identical to the collection of mammals, just as the empty set is not empty but a sign *marking* the empty set. The empty set does not exist as a difference in and for itself. It exists only as a mathematical mark that has some meaning to a mathematical community that considers it to be of some value: a difference that makes a difference. Whenever a difference is formed, like a set, for instance, it is a difference, which is in itself a difference, a difference that makes a difference. Therefore, the worldview of different organisms is determined not by differences existing in isolation from the minds representing them but through the coupling of certain observing and observed systems.

This idea of a difference that makes a difference brings us back to Whitehead, who proposed that the laws of algebra do not depend on arithmetic but "entirely" on certain conventions. The whole of mathematics, argues Whitehead, "consists in the organization of a series of aids to the *imagination* in the process of reasoning; and for this purpose *device* is piled upon *device*" [ibid., my emphasis]. Whitehead considers imagination as the ability to perceive possibilities beyond immediate reality, where reality is our given representation of knowledge. In this context, we may explain the notion of a variable by saying something like this: imagine a thing we name x, which stands for a set of possible values. "A great idea," one may say, "can it stand for a Christmas turkey?" Imagination, you explain, can be wild, but in the context of algebra, it is constrained by axioms, rules, and operations. For x in the equation $x + 1 = 2$, the only possible value is 1. Although x can signify a variety of values, this potentiality is actualized through some constraints that result in one possible solution, which is 1. The value of x actually exists even before we solve the equation. It is not the same as the potentiality of a fertilized egg being able to turn into a baby. The actuality of x exists before we solve the equation, but it materializes only when we solve it. Paradoxically, constructive imagination exists only through some constraints, another Kushiya that would probably have delighted Monsieur Chouchani.

3.5 Mind on the Fly

So far, I have explained that the basic unit of the mind is the primary distinction, a difference that encapsulates surprisingly complex properties such as self-reference. We can imagine an increasingly complex domain of abstract entities and operations through our sign system. However, a point repeatedly discussed but not fully elaborated is the dynamic nature of the mind. This is a watershed differentiating between thinkers such as Bohm, Whitehead, Bateson, and others who focus on stable entities. What does it mean to think in terms of processes and dynamics? It means that we focus on *events, interactions, and transformations*. Energy, as explained before, is defined through transformations. There is no visible object called "energy." Energy is not a thing but an idea expressing different and well-specified forms in transformation. It is somewhat surprising that, while such a foundational aspect of nature, underlying all forms of existence, must be treated dynamically, constructions of the mind are treated differently. There are good explanations for why this is the case [7], but here, we may inquire about possible alternatives and how they can shed some light on the processes of the mind and mathematics.

Let me return to David Bohm and his paper "On a new mode of description in physics" with Basil Hiley [8, p. 172], where the authors suggest that: "New informal languages and their extensions into mathematical forms need to be investigated" to give "primary relevance to activity." This approach [ibid., p. 175, my emphasis], they further argue, should give "a basic role to the *verb* (while nouns will be regarded as abstractions from verbs)."

Bohm's emphasis on the verb unknowingly echoes the centrality given to the verb by the French linguist Tesnière. There is an excellent reason to focus on the verb as the central unit of an *event-focused approach*. To explain and illustrate this idea, let us examine the sentence: John ate the cake. From an entity-based perspective, the sentence is a representation of two objects (i.e., arguments): John and the cake, associated by the verb (i.e., predicate) "ate." From an event-centered perspective, the focus should be the verb that "defines" John as the one eating the cake and the cake as something that is eaten. Forming a matrix of things eating and being eaten, we may group them into two new classes of eaters and food. Look at the following matrix associating different arguments with "eat":

	Potato	Meat
Goat	+	
Cow	+	
Lion		+
Wolf		+

Through the matrix, we can group a goat, cow, lion, and wolf into the same class of objects, as defined by their participation in the eating event. This set differs from other things not associated with active eating, such as a shoe or a book. Similarly, the columns represent two sets that can be defined as food through their participation in the eating event. Despite the difference between the sets populating the columns and the rows, they can be grouped into two novel and abstract classes. It is the relational event of eating that forms these sets. However, the similarity also paves the way for differences, as we observe that goats and cows consume potatoes, while a lion and wolf consume meat. This difference of similarities now forms a more refined matrix in which carnivores are separated from vegetarians. A network of events centered around interactions would give us a representation of the way things are related and differentiated through dynamic events such as the eating event.

The verb not only represents the event but, similarly to energy, expresses a *transformation*. The verb "ate" transformed the cake from one temporary and stable state into a different state, forming a different distinction accompanied by different conclusions. For instance, an eaten cake does not exist anymore, and one may even argue that an eaten cake is not a cake anymore. Commonsensical knowledge, such that an eaten cake cannot be eaten, is far from trivial, and we probably learn it from events as suggested above.

Transformations can form identities through repetition, for instance, or destroy them, such as in the case of the eaten cake. Here, we understand that reality is more like a network of events in which things are converted and transformed than a warehouse of entities passively lending themselves to our minds. If this is the case, a child facing the arithmetic expression $2 - 1 = 1$ is actually facing a transformation of the class name (i.e., the name of all sets having two objects) into another sign (i.e., 1) through the operation/verb of subtraction. This explains the difficulty of mathematics. Not only do we deal with abstract objects of a culturally formed domain (i.e., mathematical objects), but we also have to deal with forms and transformations that cannot be trivially grounded in our basic sensorimotor experiences. This point is further elaborated in the next section.

3.6 Mind as Matrix and Transformation

To understand how the mind may be represented as a network of sign-mediated events, we may use linear algebra and illustrate it with respect to the verb. More specifically, to represent mathematically how a verb transforms its subject and object, we can use a framework combining vector space models with transformations. This approach allows us to model the dynamic changes in a state formed by the verb.

Let me start with the idea of the vector as an ordered array of numbers. To represent the meaning of "cake," for instance, we can identify its embedding, which consists of the words that define the semantic field of "cake." For example, we know that "cake" is associated with words such as "birthday," "chocolate," "cheese," etc.

It is less associated with words such as "toilet" and "octopus." When representing the meaning of "cake" as a vector of numbers, we will use an ordered array of numbers. Each number is a weight associating the word in the vector with our target word (i.e., "cake"). The embedding, therefore, represents the semantic field of the word. For simplicity, let us represent the meaning of a cake using four dimensions (i.e., words) to cover the total semantic space and values of 1 and 0. This gives the vector:

$$\begin{bmatrix} 1 \\ 1 \\ 0 \\ 0 \end{bmatrix}$$

The first number in the vector corresponds to "birthday," the second to "chocolate," the third to "toilets," and the fourth to "octopus." This vector represents the meaning of "cake" as a point in a four-dimensional space, where the words collocated with "cake" are the coordinates defining the semantic space in which "cake" resides and its exact location.

From a static perspective, the vector is just an array of words. From an event-focused approach, the vector can be considered instructions, or a procedure, for finding "cake" in the semantic space in which it resides. To explain this point, let us assume that "cake" resides in a 2D space defined by the coordinates/dimensions: "birthday" and "chocolate." The vector defining "cake" is:

$$\begin{bmatrix} 2 \\ 2 \end{bmatrix}$$

Its representation in the coordinate system is shown in Fig. 3.1.

The vector can be interpreted as directions for finding "cake": First, start at the origin and go two steps to the right. Next, go two steps up. Representing meaning as a vector is a way of considering the sign in dynamic terms and as a list of moves directing us to a specific location. Do you remember that I quoted McCarthy in the previous chapter? He wrote [9]: "Nobody comes with names. You give them names so that you can find them in the dark." The idea of the vector as a process bringing you from one point to the next corresponds perfectly to McCarthy's insight. The meaning of the sign "cake" is represented through a vector. The vector is the name we give to "cake," which is actually the instruction for finding "cake" in the dark, the darkness of a semantic space. In this context, the sign is substitutive in the most dynamic sense, leading us through a space of possibilities. In other words, representing the sign through its embedding and as a vector can be interpreted in dynamic terms as the operation we apply to find the sign in the semantic space.

This idea can be further illustrated in the context of mathematics and multiplication. We are usually taught as young school children that *multiplication* is just addition in disguise. For instance, 2 * 3 is just 2 + 2 + 2. So, 2 * 3 can be interpreted by

Fig. 3.1 The
representation of cake in a
2D semantic space.
(Source: Author)

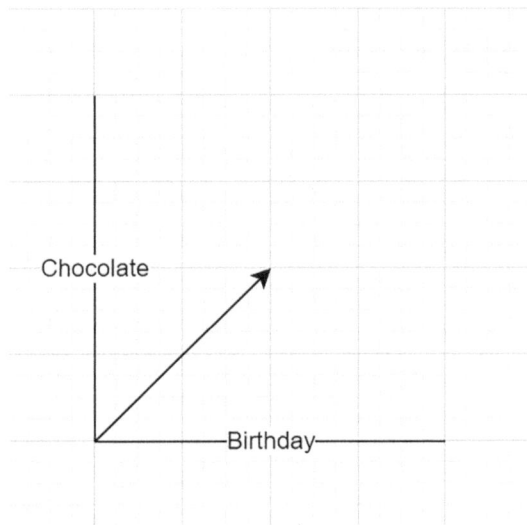

the child as 2 + 2 + 2, but when we have $\frac{1}{2}*\frac{1}{3}$, what does that mean? Where is the addition here? As argued by Devlin[1]: "I wished schoolteachers would stop telling pupils that multiplication is repeated addition." The vector representation of numbers may give us a different interpretation of multiplication. For instance, let us represent the number 2 by the vector:

$$\begin{bmatrix} 2 \\ 0 \end{bmatrix}$$

and multiply it by 2:

$$2\begin{bmatrix} 2 \\ 0 \end{bmatrix}$$

to obtain:

$$\begin{bmatrix} 4 \\ 0 \end{bmatrix}$$

This is represented visually in Fig. 3.2.

---

[1] https://www.jonathancrabtree.com/mathematics/devlin-on-mira/

Fig. 3.2 Visualizing
multiplication. (Source:
Author)

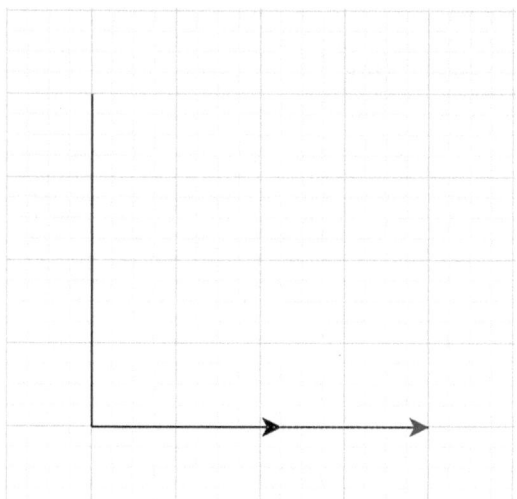

By multiplying the vector representing 2 by 2, you can see that we have *stretched* and *scaled* it on the x-axis from 2 to 4. Why is that important, and what new things can we learn about multiplication?

Teaching multiplication as addition may be justified as a heuristic appropriate for a specific developmental phase of young children. However, seeing multiplication in vector terms emphasizes a more general and meaningful interpretation of multiplication that is relevant to our minds as creatures that are highly sensitive to scaling in our environment. Multiplying a vector by a number corresponds to scaling the vector, and scaling is not a simple addition. On the contrary, in natural phenomena, scaling is NOT additive. Things do not change in size through a simple additive dynamic. Let me explain this point.

We all know that size matters and sense the environment through a specific logarithmic scaling function. When we think about our mind in processual terms, even the basic arithmetic operation of multiplication seems different, and in a way that is much more aligned with our experience as beings living in a natural order where scaling counts and has its natural logic.

So, let us adopt the proposed event-focused approach and return to John, who ate the cake. I aim to show how the verb, as a transformative activity, may change the meaning of "John." This highly focused and artificial example may be used to understand the general event-centered approach to mind and mathematics.

Let us see how the verb "eat" may transform the vector representing "John." For simplicity, we assume that the vector for "John" is a two-dimensional vector with values for "satiate" (0.3) and "man" (0.7):

$$\begin{bmatrix} 0.3 \\ 0.7 \end{bmatrix}$$

"Man" and "satiate" are two coordinates/dimensions through which the meaning of "John" is specified. We ignore the absolute meaning of the numbers. We can see that John is a man to the extent of 0.7 and satiate (i.e., not hungry) to the extent of 0.3. The two properties specify the meaning of "John." As the vector represents the meaning of "John," we are interested in how John is transformed through eating. The transformation can be defined as a 2D matrix with the following values:

$$M_{\text{EAT}} = \begin{bmatrix} 30 & 0 \\ 0 & 1 \end{bmatrix}$$

To better understand the meaning of this transformation, consider Fig. 3.3, where the unit vectors $\hat{i}$ and j represent the properties of being satiate and being a man, respectively.

Knowing that John ate something, nothing is likely to change in the property that John is a man. This is assumed to be the same. However, it is much more likely that he will become satiated (see Fig. 3.4).

Given the limitation of space, I didn't stretch the value of "satiate" to its new length. However, one can see that the transformation changes nothing in the property of being a man; it only affects the property of being satiated. How did I get the numbers for the matrix? I have chosen them to illustrate a point. However, one can model a transformation by learning from experience, even from one's own experiences. By observing yourself, you may have learned that eating can turn a hungry person into a satisfied person. Eating, though, has nothing to do with transforming one's gender.

To understand how deeply the idea of transformation is grounded in our minds, one only has to read Ovid's *Metamorphoses* or children's stories, where imaginary transformations are common. Even a child whose mind is not yet mature enough to

Fig. 3.3 The unit vectors.
(Source: Author)

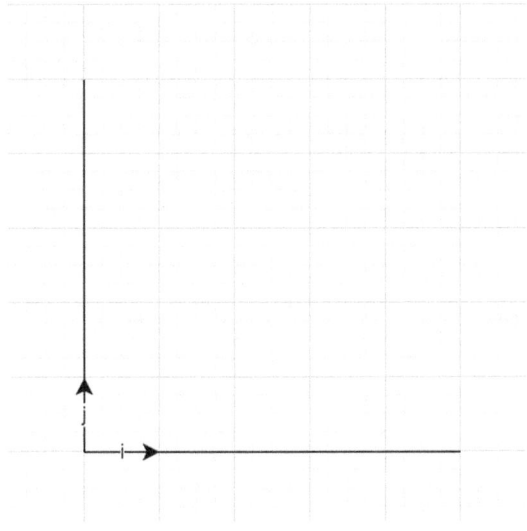

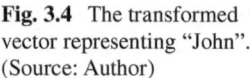

Fig. 3.4 The transformed
vector representing "John".
(Source: Author)

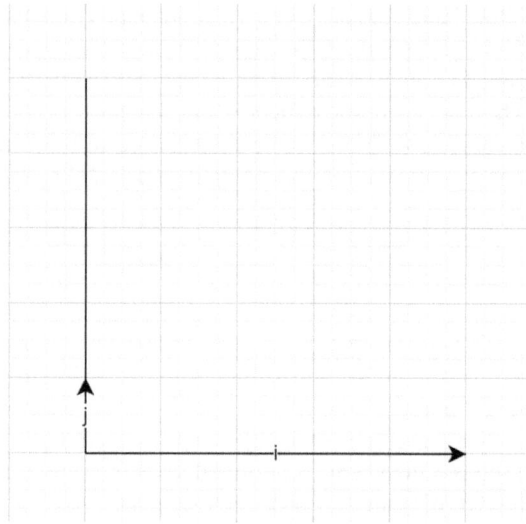

understand the most basic concepts of linear algebra will be fascinated by the story
of Cinderella, in which a pumpkin is transformed into a carriage, or the story of
Rumpelstiltskin, where straw is turned into gold. These transformations have no
grounds in reality, where a pumpkin has never been observed to turn into a carriage
or straw into gold. However, imagination, particularly the imagination during our
early developmental phases as children and as a civilization, displays some fascinating processes underlying our individual and collective minds. Now, let us return to
the matrix.

The first column of the matrix shows how the first dimension of John (i.e., satisfied) is transformed knowing that he ate:

$$0.3 = \begin{bmatrix} 30 \\ 0 \end{bmatrix} = \begin{bmatrix} 9 \\ 0 \end{bmatrix}$$

The second column of the matrix transforms the second dimension of John (i.e.,
being a man):

$$0.7 \begin{bmatrix} 0 \\ 1 \end{bmatrix} = \begin{bmatrix} 0 \\ 0.7 \end{bmatrix}$$

The combined influence of the matrix on John is calculated as follows:

$$\begin{bmatrix} 9+0 \\ 0+0.7 \end{bmatrix} = \begin{bmatrix} 9 \\ 0.7 \end{bmatrix}$$

Compare the original vector of John to the transformed vector:

$$\mathbf{v}_{\text{John}} = \begin{bmatrix} 0.3 \\ 0.7 \end{bmatrix} \mathbf{v}_{\text{John\_transformed}} = \begin{bmatrix} 9 \\ 0.7 \end{bmatrix}$$

Being a man is a property that has not been transformed by eating a cake. However, being satiated is a property that has been significantly changed. This is a critical point. Being John is an idea that cannot be exhausted by a token of stability; it is a relational structure represented by a vector transformed through eating. There is no John outside the semantic space of coordinates, and the location of John changes dynamically, evolving and unfolding as we examine it within the matrix of events. The implications of this event-centered approach for inference/reasoning cannot be underestimated. For example, we may ask whether John is hungry after eating the cake. In reality, John may still be hungry or not hungry, but given his transformation through eating, the probable answer is that he is now more likely to be satisfied. In other words, we can infer that John is more likely to be satisfied *after* eating. John gives a specific case of a more general form of reasoning: people are expected to be less hungry after eating rather than before eating. Here, we can also better understand how order is formed through transformation. John before and after eating are two asymmetric representations of John, and in fact, the before and after are defined by the asymmetry following the transformation. Therefore, we can think about order in mathematics and the mind as grounded in the asymmetry created by a transformation.

The abovementioned process of transformation can be more formally presented as follows. First, we represent each noun as a vector. For example, John is represented by the vector $\mathbf{v}_{\text{John}} \in \Theta^d$. Next, we define the matrix for EAT. This matrix represents how John's vector is transformed in the context of EAT. By applying the transformation to John, we now have a new vector $\mathbf{v}_{\text{John\_transformed}}$, which represents the way the original meaning of John has been transformed using the function defined by the matrix. It shows us that John is more likely to be satisfied after eating and that nothing has changed regarding his manhood, at least not due to eating a cake.

Previously, we discussed syllogisms governed by strict logical rules. Here, the process of reasoning is grounded in different assumptions. First, the meaning of objects does count, and it is represented through a matrix of relations between the object and the objects associated with it in the semantic space. When we try to understand who John is, we do not turn to dictionaries but instead identify the *embedding* of John. Second, the process of reasoning results from the *transformation* applied by the verb EAT. The matrix for EAT operates on the vector of John, and the transformed representation of John allows us to gain some probabilistically based conclusions grounded in the transformation process.

This event-centered approach seems to be much more similar to the way human beings actually think and the way modern tools of AI are similar to human thinking. Instead of thinking about John and the cake as stable entities floating in space, we represented them *relationally* as points in a high-dimensional space defined by coordinates. Therefore, the meaning of these two entities has been described by a relational approach. Moreover, specific emphasis was given to the verb and the way it transforms the two relational objects. The meaning of John, although represented by a vector, has been deeply related to the way certain interactions transform it. John's interaction with the cake may have transformed John from a hungry person to a satiated person. The cake has been transformed from a state of existence to a state of nonexistence. The local event in which John ate the cake may be, in itself, an event transforming others. For example, if John ate the cake his wife baked as a surprise for her mother's birthday, this event might transform John's wife in the short run, but hopefully only in the short run.

At this point, we may return to our discussion of mathematical operators. When transforming a number by multiplication, the transformation allows us a better understanding of the objects involved. For instance, Ishibashi and Moriguchi [10] studied scale error among children. This occurs when children try to perform inappropriate actions on small objects because they have not taken into account the object's actual size. For example, when children try to get into spaces that are too small, they express the scale error. Think, for instance, about the legendary story of Aladdin. In the story, the Genie gets in and out of a lamp. One amusing part of the story is probably the Genie's successful violation of scale. It is perhaps not a coincidence that children's movies include references to children's imagination expressing possibilities beyond reality.

Mathematical imagination seems to reside in the basic dynamics of the mind. This imagination is not always appreciated by adults. Imagine a group of narrow-minded educators invited to the premiere of Aladdin. When they see the Genie sneak back into the lamp, they may be outraged and protest against the film's didactic error and the way it could amplify young children's scale errors. From an event-centered perspective, one may calm them by explaining that the transformations are those defining the object. If this is the case, then our imaginary Genie does not enact a scale error but teaches us about the unique nature of Genies, who have the ability to act indifferently to the actual size of objects. The imaginative character of the Genie may also trigger some interesting scientific observations. For example, Genies may remind us of cats, which also have an amazing way of sneaking into small, confined spaces. It has been argued by a respected physicist, who won the Ig Nobel Prize for his suggestion[2] that cats are liquid [11] rather than solid matter! This conclusion should be praised for its creativity. This research shows one benefit of using an event-centered approach and imagination.

Now, back to the children. Ishibashi and Moriguchi [10] found that the most reasonable explanation for scaling errors is "children's immature conceptual representation" of size. As we age as individuals, we hopefully gain a better and better understanding of scale and the way the size of specific quantities changes with respect to other quantities. One example of scaling is expressed in *allometry* by the relation between metabolic rate and body mass. Larger animal species, such as

---

[2] https://www.pbs.org/newshour/science/answering-the-question-that-won-me-the-ig-nobel-prize-are-cats-liquid

elephants, have a lower heart rate than species with smaller body sizes, such as the mouse. The law associating these quantities is well-formulated and fascinating, and it is not an additive law. It is a law of multiplication that can be better understood in terms of scaling a vector. *There is new scope for the imagination to be explored by considering transformations in terms of scaling.

Points to Note
- **Complexity of mind**: reducing the mind to mathematical concepts is impossible.
- **Embodied mind theory**: paradoxically, it shows us that the basic units of the mind are complex, abstract processes.
- **Event-centered approach**: we should study the mind and its expression in mathematics through events, interactions, and transformations.

References

1. Halmos, P.R.: Naïve Set Theory. Springer, New York (1974)
2. Harnad, S.: To cognize is to categorize: cognition is categorization. In: Handbook of Categorization in Cognitive Science, pp. 21–54. Elsevier, New York (2017)
3. Lakoff, G., Johnson, M.: Philosophy in the Flesh—The Embodied Mind and its Challenge to Western Thought. Basic Books, New York (1999)
4. Whitehead, A.N.: A Treatise on Universal Algebra: With Applications. Cambridge University Press, Cambridge (1898)
5. Bateson, G.: Steps to an Ecology of Mind. University of Chicago Press, Chicago (2000)
6. Neuman, Y.: Mathematical Structures of Natural Intelligence. Springer, New York (2017)
7. Neuman, Y.: Processes and Boundaries of the Mind: Extending the Limit Line. Kluwer, New York (2013)
8. Bohm, D., Hiley, B.J.: On a new mode of description in physics. Int. J. Theor. Phys. **3**(3), 171–183 (1970)
9. McCarthy, C.: Stella Maris. Alfred A. Knopf, New York (2022)
10. Ishibashi, M., Moriguchi, Y.: Understanding why children commit scale errors: scale error and its relation to action planning and inhibitory control, and the concept of size. Front. Psychol. **8**, 251336 (2017)
11. Fardin, M.A.: On the rheology of cats. Rheology Bulletin. **83**(2), 16–17 (2014)

Chapter 4
Symmetry, the Unconscious, and Imagination

Abstract Symmetry is expressed in mathematics, psychology, and the arts. In this chapter, I inquire about the symmetry dynamic of the unconscious, as explained by Matte-Blanco, and point to the importance of this mind system in supporting imagination. We will meet aliens at a galactic family gathering and the Frog Prince, and see how projection matrices can model the way our minds move between different dimensionalities.

4.1 From Dada to Freud

Johannes Theodor Baargeld (Alfred Emanuel Ferdinand Gruenwald) was a painter associated with the Dada artistic movement. His painting Beetles[1] (1920) presents sketches of beetles in four parallel rows. Four beetles appear in each row, and each beetle is a transformed or deformed version of the others. The last two rows of beetles are presented as if they appear on a lead sheet. In music, a lead sheet includes symbols, such as notes, providing the essential information for musicians to generate a piece of music. When we look at Beetles, just for a moment, our minds may unconsciously mistake the beetles for musical notes. They are just similar enough for our minds to consider them identical and ask what music might be played if a musician could actually play them. The similarity of the beetles and musical notes is part of the aesthetic experience evoked by the painting, and so is the musement of imagining what it sounds like to play beetles.

The painting presents the theme of symmetry. Each beetle seems to be produced by a simple algorithm generating a symmetric pattern through repetition. Starting from the top-left position of the painting, a beetle is painted. Then, a copy is created and located at a fixed distance from the previous beetle. We move to the following line and repeat the process. However, the beetles are not perfect reproductions of each other. Each beetle is a transformed/deformed version of a prototypal beetle.

---

[1] https://www.moma.org/collection/works/35886

Some images of the beetles don't look like beetles at all and can be interpreted as beetles only because we see them in context. What we observe is not perfect symmetry but a variation of symmetry. This exciting pattern is formed through repetition with variation.

Repetition with variation [1] is a procedure known to create an aesthetic experience. To illustrate this point, I asked the artificial intelligence Dall to recreate the painting using abstract symbols and forms corresponding to Beetles. It generated the image in Fig. 4.1, which I decided to call "Aliens in a galactic family gathering."

This image reminds me of paintings by Joan Miró and wonderfully expresses the aesthetic value of symmetry. Again, this is not perfect symmetry, but symmetry with variations. The image in Fig. 4.1 does not look the same as the symmetry of a sunflower or even the symmetry that appears in Baargeld's painting. It looks like a painting of a family of aliens having their picture taken during a galactic gathering. I hypothesize that what may excite us in the painting shown in Fig. 4.1 is the interplay of similarity and difference, symmetry and asymmetry working in concert with the optimal amount of injected noise to turn the painting into something alive. It is not a chaotic spread of colored dots, as seen in one of Jackson Pollock's paintings, nor is it a perfect mathematical symmetry that can be formalized through abstract algebra. Here, we have a sophisticated structure with elements expressing similarities and differences.

Surprisingly, it has been argued that this interplay of symmetry and asymmetry is expressed by our mind. Matte-Blanco (1908–1995) was a psychoanalyst who

Fig. 4.1 Aliens in a galactic family gathering. (This image was created with the assistance of DALL·E 2)

proposed a fascinating "mathematical" theory of the unconscious, which is grounded in his Principle of Symmetry (PS). It was argued by Matte-Blanco [2, 3] that our thinking is governed by two systems working in concert. While our rational thinking is governed by asymmetry, the unconscious (Unc) part of our mind expresses the logic of symmetry. To explain this fascinating and challenging theory, we must first turn to Freud and his idea of the unconscious [4].

As introduced by Freud, the concept of the Unc suggests that some mental aspects of our life exist outside our awareness. This aspect is self-explanatory to anyone who has reflected on their thoughts and behavior. However, Freud justified the concept by describing it as a *necessary assumption* [ibid., p. 166]. It is necessary, explains Freud, because our conscious life includes too many gaps that cannot be described in terms of conscious processes. As a Jew, Freud may have noticed a Kushiya and sought to resolve the Kushiya of the conscious by filling the gaps through the Unc. Let me illustrate this Kushiya. Attending an academic conference, I observed a young student obsessively washing his hands in the men's room. Waiting for my turn to use the basin, I noticed that he was doing so with no intention of completing the task quickly. He was obsessed with washing his hands with no rational justification. How can we explain this obsessive-compulsive behavior? For Freud, the fact that the person cannot reasonably explain his behavior is a clear justification for assuming that the explanation resides in a realm different from the agent's consciousness. This is the realm of the Unc.

Here is another example: Children's stories provide a great source of amusement, but for reasons one cannot easily understand from a "rational" perspective. Think, for example, about *The Frog Prince*, a story where a prince has been transformed into an ugly frog by a vicious witch and is saved only when the princess kisses him. One has to admit that this is a strange story. A prince transformed into a frog, then retransformed into a prince? This transformation is impossible from a rational and scientific perspective. First, no one has ever observed a transformation where an individual being gets transformed into a different thing, whether a frog, a cat, or a spoon. Second, transformations that involve complex living systems, such as princes and frogs, are asymmetric and irreversible. While water can turn into ice and vice versa, a caterpillar turning into a butterfly presents the irreversible transformation of a complex living system. The ultimate example of an irreversible process is that of dying. A living person is transformed into a corpse, as inevitably happens to all of us. No one has ever observed the reverse process where the dead returned to life. Death is irreversible. What about in mathematics? There, we have the idea of inverse objects. For example, for any number x in the set of real numbers, the additive inverse is $-x$, because $x + (-x) = 0$. Here, 0 is the identity element for addition. But reality is irreversible, so how can mathematics use such an unrealistic inverse? Like children's stories, mathematics is grounded in imagination and the logic of the unconscious.

In this context of asymmetric transformations, the Frog Prince story is irrational. It sounds as if it were written by a madman who knows nothing about reality. Some people might even be outraged that such a story could be published and told to young children. In fact, in his book "Émile, or On Education," Jean-Jacques

Rousseau, the famous eighteenth-century philosopher, criticized fairy tales and other forms of imaginative literature for children, arguing that they promote unrealistic thinking and are not conducive to the moral and rational development of young minds. Well, Rousseau's *moral* criticism deserves specific attention. Like many other philosophers, he probably did not know much about children's minds, feelings, emotions, and morality. We know that Rousseau was the father of five children who were sent to the orphanage. Nice, isn't it? The philosopher of the Enlightenment, the "enlightened lighthouse" who criticized fairytales for corrupting the minds of young children, sent his children to an orphanage without considering the pain children bear when taken away from their mothers and raised by strangers. This anecdote is only one instance where the claim of being rational and enlightened is accompanied by the opposite result and its consequences.

To explain the magic of children's stories, one must assume a mechanism beyond consciousness. When children are fascinated by fairy tales, there is a good reason for their excitement, which cannot be justified "rationally." This is why Freud considered it essential to assume the existence of the Unc. The Unc is not a randomly generated aspect of our mental life. Freud's real achievement was probably not just pointing to the Unc, but identifying its characteristics and the way it works. For example, one characteristic Freud identified is the absence of time in the Unc. The Unc is *timeless* [ibid., p. 186], and events "are not ordered temporally, are not altered by the passage of time; they have no reference to time at all."

Previously, we learned that order is a fundamental cognitive and mathematical concept. The idea of order is a cornerstone of the rational mind. However, as argued by Freud, it has no place in the Unc. For example, in *Alice's Adventures in Wonderland*, Alice grows and shrinks in a way that would be impossible by any natural growth process, since such processes are always irreversible and asymmetric. The order of things in Alice is irrational and probably expresses the underlying dynamics of the Unc. While the growth of little girls is irreversible, the imagination expressed in the story uses an important resource of the Unc, which is not obliged to respect any sense of order.

Matte-Blanco [2, 3] proposed that the complex and bizarre dynamics of the Unc can be theorized through two main principles integrated into a system he described as *bi-logic*. The first is *the Principle of Generalization* (PG), where mental elements may be considered in terms of higher and higher collections of which they are a part. For example, [2, p. 45] tells us about a patient who said MB's assistant was rich. When asked to explain herself, she replied, "He is very tall." MB explains her irrational answer as follows: the assistant is tall. The property of being tall is a subclass of a higher-level category of things described as *high*. A rich person has a high salary; both rich and tall are included in the same category of *high*. Therefore, tall is high, high is tall, rich is high, high is rich, and at the class level, being tall and rich can be exchanged. *The part is identical to the whole, and the parts of the whole are identical.*

Such a confusion between levels of analysis has been banned from mathematics, specifically after the discovery of *Russell's paradox*, but the Unc has no interest in mathematics and its developments. It is a beast that, luckily, cannot be tamed. Why luckily? Because imagination is possible only through the wild logic of the Unc. Consider, for example, the metaphorical use of "sweet." When I describe my cat Kitmon as sweet, I don't mean it is literally sweet. I have never tasted my cat and have no plans to try. I transfer the meaning of sweet from the pleasurable taste of some foods, such as honey or chocolate, into a different domain. Another example appears in a poem by the Israeli poet Yona Wallach. The poem opens with the following line: "I will never hear God's sweet voice." Of course not, the arrogant Rousseau would have said. Those who believe that they hear the voice of God should be hospitalized for insanity, and if you think that God's voice is "sweet," then you are a nut, because sweet is a taste and a voice has no taste.

We understand that sweet things compose a set, a subset of pleasurable things. My cat gives me pleasure; therefore, I describe it as "sweet," and the same holds concerning God's voice in the poem. MB explains that the symmetric logic of the Unc and the asymmetric logic of the conscious and rational parts of our mind go hand in hand. This idea has important implications for understanding mathematics. Mathematics may epitomize the idea of rational thinking, but it is impossible without the "irrational" logic of the Unc.

This exchange of a part for a whole can also be illustrated through the vulgar term "asshole." "Why use a vulgar term for a description?" a snowflake student may ask. "Don't you think using this vulgar language may hurt the feelings of some readers?" From the etymology of "vulgar,"[2] we find that it refers to ordinary, everyday people. If you are interested in an accurate understanding of the human mind, you must understand how common folk, like me and you, think and behave. The alternative is to study the mind from a detached, sterile, and invalid academic perspective, running laboratory experiments on groups of college students. In that case, your research must obey strict ethical protocols, and since psychology is characterized by poor scientific standards and heroes of nothing, like Dan Arieli, you may say goodbye to any accurate understanding of human minds.

But let's get back to assholes—and please don't confuse this with the previously mentioned academic, who has published an outstanding and highly cited psychology paper showing how recalling the Ten Commandments can have a statistically significant impact on people's moral behavior. The term "asshole" is not used literally. It is a *synecdoche*, a rhetorical form in which a part refers to the whole (*pars pro toto*) or vice versa. What does it mean when we describe someone as an asshole? It is an "irrational" description because the whole person cannot be equated with one of his parts. However, the *substitutive* logic of the Unc suggests the following process. A certain person evokes strong negative feelings in us through his behavior. The anus (i.e., the asshole) is considered an area associated with negative attributes such as dirt, feces, and bad smells. Negative emotions may create an association between the whole person and the part, allowing us to synchronize through *synecdoche*. When we say "Roger is an Asshole," we point out a similarity between

---

[2] https://www.etymonline.com/search?q=vulgar&utm_campaign=sd&utm_medium=serp&utm_source=ds_search

two different things that exist at two different levels of organization: the part and the whole. "Paradoxically," the rhetorical power of the synecdoche may gain its effect by being irrational.

The Principle of Symmetry (PS) suggests that in the Unc, relations can be reversed. For example, if John is the brother of David, then David is the brother of John. The relation of brotherhood is symmetric. However, if John is the father of David, then the relation is *asymmetric*, and it is false to say that David is the father of John. MB argued that the Unc is governed by the symmetrization of asymmetric relations. Take, for instance, the classic movie *Back to the Future*. Marty, the film's teenage hero, flies back in time using a time machine. The symmetrization of time is impossible in reality, but it is in the movie, which is an expression of the Unc's dynamics in artistic form. Going back in time, Marty's adventures become complicated when his future mother, Lorraine, falls in love with him. This twist in the plot is of dramatic power, as one can imagine the sin of incest and the possibility of Lorraine having sex with her son. However, if Lorraine does not marry Marty's father, then Marty cannot be her son, even though he is!

This situation sounds like a typical case for a Talmudic discussion of the future. Imagine two Talmudic scholars in the year 2300 being informed about a time machine. They are told that a guy named Marty returns to the past and falls in love with his future mother, Lorraine. The scholars are asked whether having sex with Lorraine is considered the unforgivable sin of incest. "This is a very difficult Kushiya," reply the scholars. "On the one hand, Lorraine is Marty's mother. Therefore, having sex with her is forbidden. On the other hand, when Marty meets Lorraine in the past, she is not his mother yet. Therefore, it cannot be called a sin." Here, we can see how the Principle of Symmetry violates the law of contradiction. The two following propositions cannot be true simultaneously:

1. Lorraine is Marty's mother.
2. Lorraine is not Marty's mother.

The confusion resulting from the symmetrization process has a clear dramatic dimension that corresponds to the profound logic of the Unc. The idea that p and not-p can hold simultaneously is expressed in many other irrational artistic forms. For example, in the 1960s horror film *Night of the Living Dead*, we meet a *zombie*. A zombie is something that is both dead and alive. It is an imaginary creature, as a person cannot be dead and alive. A person is always either dead or alive.

The PS may also be expressed in the confusion between the representation and the represented. In one of his anecdotes, Bateson describes a madman eating the menu because he has confused the menu for the food. The menu is not the food, just as a map is not the territory it represents. This confusion also expresses another property of the Unc: the confusion between internal and external reality. The representation is an internal reality. It is not reality. Bateson continually reminds us that the map is not the territory. But the map can be mistaken for the territory whenever the PS is applied.

4.2   Mathematical Imagination and the Unc

The idea that the Unc is governed by the PS is of great interest to us, as it expresses the idea of similarity discussed previously. A cat and a pigeon are both animals; therefore, they are similar. However, they are not identical, and their difference cannot be erased by knowing they belong to the same class. The logic of the Unc can treat similarity as identity. The Unc seems to be a system that drops some of the most fundamental constraints necessary for rational thought, as expressed in mathematics.

Following Freud, MB explains that the Unc and the conscious systems are two legitimate interacting facets of any given mind, which is why he uses the term bi-logic. One important lesson in mathematics is that *imagination is possible when some constraints are relaxed, allowing us to explore different possibilities.* As will be argued in Chap. 5, this is precisely how mathematics has progressed historically.

As Tom Waits writes in his song *Innocent When You Dream* (1978): "You're innocent when you dream." We are innocent, which means we are not guilty. We cannot be guilty in a place where no laws hold. Take, for example, Alice in one of her adventures. Drinking from a bottle, what is the most likely transformation she might have experienced? I asked ChatGPT the following question: if x is a person drinking a tasty drink from a bottle, what are the three most likely things that can happen to him? ChatGPT suggested that x may feel energized, refreshed, and satisfied. However, the transformation experienced by Alice was rather different: "What a curious feeling!" said Alice, "I must be shutting up like a telescope." Alice is shrinking as a result of drinking from the mysterious bottle.

I previously argued that reasoning is deeply associated with transformations and their resulting vectors. In a system where the logic of reasoning is relaxed, unexpected, surprising, and often amusing consequences may be entailed by a given set of premises. Why is this? Because the transformations used by the Unc are followed by results and consequences that differ from those entailed by the rational transformations of the Unc. In the realm of the Unc, no one is guilty of breaking the laws of logical thinking, as there is no rational thinking, but rather another form of thinking with its own logic.

The following section aims to explore the idea of transformations in the Unc. Specifically, I would like to show how we can model the Unc process of *condensation* through transformations reducing the dimensionality of two different things. I illustrate my point through a specific example involving the similarity between the cultural representations of God and the father.

4.3   Condensing God and Father

All of us have fathers, at least biologically. Our private fathers, though, should not be mistaken for the concept of the father developed culturally in our societies. *The Historical Thesaurus* of the Oxford English Dictionary[3] teaches us that "father" was used in Old English to describe: "A person, esp. a ruler or superior, who provides protecting care like that of a father." The main property associated with the father is protecting care. However, this is not the only meaning of "father" in Old English. The sign "father" was also used to refer to Christ or the Lord. The PG then applies this to "people of God," such as priests. This dynamic is clearly expressed by the phrase opening the Catholic sacrament of confession: "Bless me, Father, because I have sinned." The father/priest is asked to forgive the sinner because he represents God, the ultimate father. Father and God have been equated in the collective mind, at least the "mind" of Christian European civilization.

We may explain this process from a psychoanalytic perspective, arguing that these two concepts have been *condensed through symmetrization*. It must be emphasized that the father equated and identified with God is not a particular father but a sign indicating a class. It has been described as the *symbolic father* that shares a common denominator with God. The symbolic God may be seen as a projection of the symbolic father, as they share several properties. For example, both the symbolic father and God represent the ultimate *authority*. The symbolic father often stands as the representative of law, order, and societal norms in psychoanalytic and cultural contexts. At the same time, God is the supreme being and moral authority in religious contexts. Another common denominator concerns protection and punishment: both figures are associated with protection and punishment. The symbolic father protects by enforcing laws and norms but punishes when his norms are violated. Similarly, God offers protection and guidance and punishes human beings for violating his law. It should now be clear why the two concepts have been condensed in our collective mind.

In Freud's psychoanalytic theory, condensation refers to the merging of ideas or images into a single sign in the unconscious mind. Concepts are not arbitrarily merged, but through a common denominator where similarity is established, resulting in identity. Is it possible to illustrate this process of condensation mathematically? Let us return to word embedding and represent the concepts of father and God using three properties: authority, punishment, and eternality.

The condensation of father and God in our minds may be described by a process in which we project these two concepts, represented as vectors, into a lower-dimensional space where their shared characteristics are emphasized. This is another way of thinking about the formation of similarities and differences. *Similarities may be formed up to identity when two different signs are collapsed into a lower dimensional space.* For example, we may project father and God, represented in a 3D

---

[3] https://www.oed.com/search/advanced/HistoricalThesaurus?textTermText0=father&dateOfUseF irstUse=false&page=1&sortOption=DateOldFirst

space, into a 2D space, into a 1D space (i.e., a line), and even into a 0D space (i.e., a point), where they are identical because all differences are lost. To illustrate such a process, let me start with two vectors whose components express the extent to which each property is associated with the concept:

$$
v_{\text{father}} = \begin{bmatrix} 0.5 \\ 2 \\ 0.5 \end{bmatrix} \quad v_{\text{God}} = \begin{bmatrix} 1 \\ 4 \\ 1 \end{bmatrix}
$$

We see that each vector has three rows representing the three properties on which it is defined. These are the dimensions of authority, punishment, and eternality. For instructional reasons, I used numbers describing God's properties as amplifying the father's properties by a factor of two. It is as if God is a scaled form of father. For example, God is twice as eternal as father. This is not, of course, a valid representation, as the property of being eternal is binary (1 or 0). However, I give precedence here to ease of presentation.

Next, I use a transformation matrix to project the vectors into a lower dimensionality. Matrices are important in linear algebra as they are used to transform vectors. Here, I use a matrix to transform and reduce the dimensionality of the vectors. For example, here is a projection matrix that I use to reduce the dimensionality of God and father:

$$
\mathbf{P} = \begin{bmatrix} 1\,0\,1 \\ 0\,1\,2 \end{bmatrix}
$$

You can see that the matrix has two rows and three columns. It is a two-by-three matrix. We start with three dimensions and project them into a 2D space. In contrast, the following projection matrix has three rows and two columns:

$$
\mathbf{P} = \begin{bmatrix} 2\,3 \\ 7\,1 \\ 8\,9 \end{bmatrix}
$$

This matrix represents a transformation from a 2D to a 3D space, where columns 1 and 2 represent the places where the two basis vectors of the 2D space end up. Projecting the father vector, we apply the following transformation:

$$
\begin{bmatrix} 1\,0\,1 \\ 0\,1\,2 \end{bmatrix} \begin{bmatrix} 0.5 \\ 2 \\ 0.5 \end{bmatrix} = \begin{bmatrix} 1 \\ 3 \end{bmatrix}
$$

And for God:

$$\begin{bmatrix} 101 \\ 012 \end{bmatrix} \begin{bmatrix} 1 \\ 4 \\ 1 \end{bmatrix} = \begin{bmatrix} 2 \\ 6 \end{bmatrix}$$

The transformed vectors of father and God are linearly dependent, and therefore, the determinant (det) of the matrix:

$$A = \begin{bmatrix} 12 \\ 36 \end{bmatrix}$$

is

$$\det(A) = 0$$

In other words, by examining the determinant of the matrix formed by the transformed vectors, we confirm that the vectors are linearly dependent (since the determinant is zero). This linear dependence in the transformed space reflects the condensation process, where shared characteristics of the original vectors are highlighted and merged into a more straightforward representation.

Mathematically, the linear dependence of the transformed vectors means they lie on the same line in the 2D space. The transformation has thus condensed the original vectors in the 3D space into a lower-dimensional space where their shared characteristics are emphasized. Put another way, we can use the transformation matrix to project the complex, multidimensional meanings of "father" and "God" into a simpler space. This is, of course, a toy example. In reality, we may want to reduce linearly independent vectors. For such a case, different approaches are required.

What lessons may be learned by modeling condensation using linear algebra? First, note that although we have reduced the dimensionality of the two concepts, we could also have projected them onto the same point, forcing them to be the same. Going from similarity to identity may be modeled as projecting vectors into lower and lower dimensionalities in which their similarity is emphasized. This means that, instead of using the noun "similarity," we can use the verb "similarization" to highlight the process through which different things can be expressed as more and more of the same by adopting perspectives of lower and lower dimensionality of their mental space. The same holds for increasing the space in which two concepts reside, emphasizing their differences.

In sum, the idea of similarity can be expressed through dimensionality reduction, relying on some shared properties highlighted by the transformation. This point is of great interest because it allows us to identify the properties selected in the cultural evolution of a concept when similarizing concepts from two different domains. For example, how did the priest become a "father"? Examining the Historical Thesaurus, I found the first link in the idea of grith-priest from 1391. It was a "priest who ministered to those who took sanctuary." In medieval England, the concept of sanctuary was a legal provision that allowed individuals accused of crimes to seek refuge in a church or monastery. Once within the sanctuary, the individual was protected from arrest for a specific period. During this time, they could either confess their crime and agree to leave the country or negotiate for a pardon. In this context, the priest served a function similar to that of the father: protective care. Old English proposed the idea of the father as an agent providing protective care. Later, when priests provided a similar function, the concepts of "father" and "priest" were projected into a 1D vector, a line representation of meaning, in which they shared the property of protective care. I find this fascinating, as it provides an articulated logic for the processes leading to similarity and identity.

4.4 Repetition, Symmetrization, and Dimensionality Reduction

So far, I have discussed the concept of symmetry and shown that the Unc expresses it and that similar processes underlying the transition to similarity in math are found in other expressions of the human mind. Moreover, I pointed out the importance of imagination and associated it with the Unc and processes of symmetrization and similarization, which are possible through projection into lower dimensionalities. However, MB made another important point that is closely associated with a concept discussed previously: repetition. The idea is simple. We observe repetition when a structure with n dimensions is reduced to a lower dimension. Figure 4.2 is a simple visual representation illustrating this point.

You can see that the triangle exists in a 2D space on the plane. When we unfold the triangle and project it onto a 1D space (i.e., a line), a repetition of A appears.

Fig. 4.2 Repetition resulting from dimensionality reduction. (Source: Author)

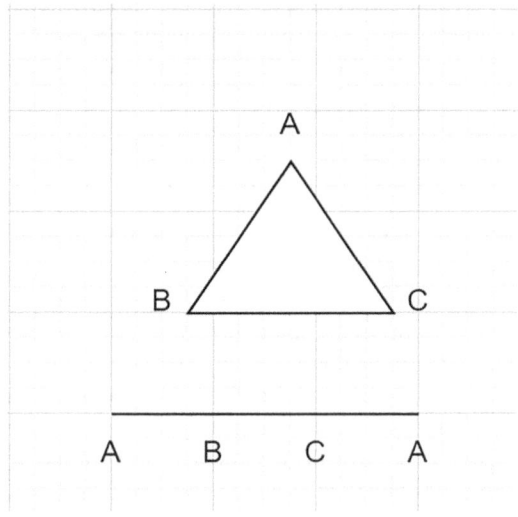

Previously, I discussed projection into lower dimensionalities as leading to *similarity* and eventually identity. We now learn that repetition is an indicator of dimensionality reduction and similarity. This is not just a bizarre idea proposed by a psychoanalyst, but a scientifically and mathematically grounded idea. Recurrence quantification analysis (RQA) is a highly productive approach for analyzing dynamic systems. It considers repetitions in a time series as signs used to reconstruct the signal in a higher dimensional space. Coming back to repetition in the present context, we may ask whether the idea can be applied to the expression of the human mind in other domains.

Let me return to poetry and, in particular, to Dylan Thomas's poem "Do not go gentle into that good night." I want to examine the words "light" and "night," which, by rhyming, form an association and hence a repetition. In the literal sense, "night" is used in Old English to describe "the time at which darkness comes on; the close or end of daylight; nightfall," and light refers to "a source of illumination." If we consider night/light as a repetition, we may ask whether a higher-dimensional space could be reconstructed. We would have to add more dimensions to the system to construct this higher-dimensional space. Assuming a 1D poem with a single repetition, we can try to construct a 2D space. Using the same approach, I looked for a common property, or dimension, shared by light and night. In Old English, "light" appears as "an epithet or description of God, as the source of divine light," while "night" (from 1395) appears as "darkness personified, esp. as a female being or *deity*." Here, something interesting can be observed. First, let us use the literal sense of the signs. "Light" and "night" can be defined by two properties: illumination and darkness. Here are the 2D vectors for these words:

$$a = v_{\text{Night}} = \begin{bmatrix} 1 \\ 0 \end{bmatrix} b = v_{\text{Light}} = \begin{bmatrix} 0 \\ 1 \end{bmatrix}$$

Let us think about these vectors in terms of their "cosine similarity." Cosine similarity measures the cosine of the angle between two vectors, quantifying how similar the vectors are in terms of their direction, regardless of their magnitudes:

$$\text{cosine similarity} = \cos(\theta) = \frac{A \cdot B}{|A||B|}$$

where *θ is the angle between the vectors A and B, A·B is the dot product, and |A| and |B| are the magnitudes of the vectors.*

The cosine similarity value ranges from −1 to 1. A value of 1 indicates that the vectors point in the same direction (angle of 0 degrees), while a value of 0 indicates that they are orthogonal (angle of 90 degrees), and a value of −1 indicates that the vectors point in opposite directions (angle of 180 degrees).

What is the cosine similarity of our vectors? The cosine similarity between the vectors is 0. How do we know? We compute something which is known as the dot product between the vectors and find that it equals zero:

$$a \cdot b = 0$$

The angle between vectors refers to the smallest angle formed by the vectors when they originate from the same point. It is measured in degrees or radians. The cosine of this angle gives the cosine similarity. In our case, the angle is 90°. This indicates that the vectors are orthogonal (perpendicular to each other). Interestingly, although they are opposites, the vectors for "night" and "light" do not point in opposite directions but are *orthogonal*. What happens if we increase the system's dimensionality and include the dimension of deity? In this case, we have the vectors:

$$a = v_{\text{Night}} = \begin{bmatrix} 1 \\ 0 \\ 1 \end{bmatrix} b = v_{\text{Light}} = \begin{bmatrix} 0 \\ 1 \\ 1 \end{bmatrix}$$

and their dot product is now 1. The cosine similarity is given by:

$$\cos(\theta) = \frac{A \cdot B}{|A||B|}$$

Substituting the values:

$$\cos(\theta) = \frac{1}{\sqrt{1/2}\sqrt{1/2}} \frac{1}{2} = 0.5$$

Regarding the angle between the vectors, we obtain:

$$\theta = \cos^{-1}(0.5) = 60°$$

We see that the two vectors that were orthogonal to each other got closer by projecting them into 3D. When adding dimensions, we can increase or decrease the similarity of the vectors to reduce or create differences. How do we know which to choose? In the case of "light" and "night" and following MB, I hypothesize that their repetition signals a structure that exists in a higher dimension. I did some detective work, starting from the simple denotative meaning of the words' origins. Next, I looked for a more abstract dimension that arrived later and could add complexity to their meaning.

Let us interpret the first three lines of the poem in light of the repetition and the way it led us to the higher-dimensional representation of "light" and "night." First, "night" derives from the Old English *niht* (or *neaht*), which is related to the Proto-Germanic *nahts* and the Proto-Indo-European *nokt-*. "Night" has connotations of darkness and the unknown and is often associated with death or the end in many cultures. On the other hand, "light" derives from the Old English *leoht*, which comes from the Proto-Germanic *leuhtam* and the Proto-Indo-European *leuk-*. "Light" is often associated with clarity, life, and knowledge. It is commonly used in religious texts to symbolize the divine or the good. While representing opposites (darkness and illumination), both words carry a sense of something divine or ultimate. Many cultures and religions depict deities associated with light (sun gods, the light of heaven) and darkness (night gods, the underworld).

When we read the poet urging his father, "Do not go gentle into that good night," we may now understand it in light of the new dimension. Here, "night" symbolizes death or the end of life. The adjective "good" suggests an inevitable and natural end, but Thomas implores the reader not to accept it passively. Considering "night" with a sense of deity could imply that death is a return to a divine state or a meeting with the divine. The line can then be interpreted as a call to resist passively slipping toward this divine end, emphasizing the human spirit and willpower against the inevitable. Urging his father, "Rage, rage against the dying of the light," we may now read it as follows: the light may represent life and the divine spark within humans. The light symbolizes not just life, but the divine essence that animates it. The command to "rage" against dying may be interpreted as a powerful insistence on the value and sacredness of life itself.

Considering the etymological and symbolic connections with "deity," the poem's opening lines can be interpreted as a passionate plea for the human spirit to defy the natural dying process. The night, although a return to the divine, is not something to be embraced without a fight. Instead, the life (light) within us, which is also divine, deserves to be fiercely defended. This resistance highlights the preciousness and sanctity of human existence, the divine light within, and the struggle to hold onto it against the encroaching divine night.

This interpretation seems reasonable enough when we recall Thomas's famous line: "And death shall have no dominion." The title of this beautiful poem is taken from the New Testament, where Paul discusses the resurrection of the dead and the victory over death through Jesus Christ. The resurrection of the dead is the ultimate fantasy of reversibility. It is the event where our Unc has the upper hand over death. Deeply motivated by this image, Thomas portrays a poetic description where *raging against death and rebelling against the dying of the light is paradoxically possible by accepting the divine residing in both.* This interpretation is an exercise in imagination and no more, but one must admit that it is an exciting interpretation beyond the poem's simple and standard interpretations. It emerged through the concepts I have previously used to identify structures and processes underlying different expressions of the mind, whether in mathematics or poetry. Similarity, differences, repetition, and dimensionality form a nexus between mind and mathematics.

Points to Note
- **Repetition with variation**: creates aesthetic experiences.
- **The symmetry of the unconscious**: describes the unique logic of the unconscious.
- **Imagination in mathematics**: is grounded in the unconscious, like children's stories.

- **Rationality and the unconscious**: mathematics requires "irrational" unconscious logic.
- **Repetition and dimensionality reduction**: repetition can be viewed as dimensionality reduction.

References

1. Alexander, C.: The Nature of Order. Center for Environmental Structure, Berkeley (1980)
2. Blanco, I.M.: The Unconscious as Infinite Sets: An Essay in Bi-Logic. Gerald Duckworth and Company Limited, London (1975)
3. Matte-Blanco, I.: Thinking, Feeling, and Being. Routledge, London (2003)
4. Freud, S.: The unconscious. J. Nerv. Ment. Dis. **56**(3), 291–294 (1922)

Chapter 5
Imagination, Mathematics, and Mysticism

Abstract This concluding chapter takes us on a journey from poetry to mysticism. It opens by showing how four ancient forms of interpretation can be used to exercise imagination. It then goes on to show how complex numbers added another possible type of number to algebra and explains how this may help us deal with a Kushiya arising from Boole's fundamental law of thought. The chapter concludes by reflecting on the intersection of mind, mathematics, and mysticism.

5.1 Elliot's River

In the previous chapter, I discussed the symmetrization of objects through the symmetrization of properties and relations. In this chapter, I would like to discuss the importance of *relations* by analyzing a poem by T.S. Elliot. This is only a device for discussing the main focus of this chapter: imagination.

Let me begin by clarifying the idea of relation. A child noticing that he got a *smaller* piece of cake than his brother might experience a feeling of injustice. This sense of injustice is possible only because the relation "bigger_than" or "smaller_than" exists in the child's mind. This relation is comparative and involves the visible size of two objects. The comparison is not quantitative yet; the child cannot yet understand the question: how much bigger is your brother's cake than yours? This relation is a difference that makes a difference, involving two objects that can be compared visually by the child.

While the two objects have a visual appearance, their relation is abstract. Developing the concept of a relation, the child moves from intuition to conceptualization, as she can actively use and signify the relation BIGGER_THAN. This relation is abstract in that it can be signified with no reference to specific objects, but merely to two variables actualized in the comparison process: xRy, where R represents the relation "bigger than." Notice how the human mind expresses abstract thinking from its early developmental phases. By comparing, the child uses a relation where the variables may take different values. This can be seen as the seed of algebra because the relation is defined for variables. The *algebra of relations* is a basic and crucial aspect of our mind. Without an algebra of relations, we would be doomed to live in a world of ultimate particularity.

Peirce developed a comprehensive theory of relations[1] that categorizes them in terms of the number of entities involved. His theory is echoed in the work of Tesnière [1] and is worth discussing. Peirce distinguishes between monadic, dyadic, and triadic relations. Here is an explanation of each of these types.

A monadic relation involves only one object. It is also known as a unary relation. These are properties or attributes that apply to a single object/argument. Monadic relations are the simplest form of relation. For example, "The apple is red." The attribute or the property "red" is assigned to the object apple and can be represented as red (apple). The relation can be applied to any variable with this property; therefore, we may signify it as $R(x)$, where R = red. This does not look like a relation, as a relation is minimally defined for two objects. However, one may consider it as an intersection between the set of things X and the set of colors Y: a morphism from the set of things to the set of colors. The monadic relation is also expressed in mathematics. For instance, it may be described as a property of a number, such as the property of being an odd number. An odd number is meaningful as long as it can be contrasted with an even number.

A dyadic relation involves two objects. It is also known as a binary relation. For example, love is a dyadic relation. To love, one must love somebody else. Even if someone is the ultimate narcissist who is in love with himself, love relates that person to another, which is himself. The dyadic relation can be signified by $R(x,y)$, where x and y form an ordered pair from a given set A. For example, "less than" is a dyadic relation in mathematics. The fact that the relation is abstract, in that it can be applied to variables, is a property of the mind that is necessary for mathematics and language. When Johan says to Mariana, "I love you less than I loved you in the past," he makes a metaphorical comparison as it concerns an abstract entity (i.e., love) and a relation originally referring to quantities grounded in sensorimotor experience.

A triadic relation involves three objects. These are more complex and can represent interactions or relationships among three subjects. For example, the relation "sold" involves three objects: the seller, the sold object, and the buyer. It can be signified as $R(x, y, z)$. In mathematics, this relation is expressed by the template $x + y = z$.

Why are relations important? As I explained previously, focusing on relations and processes may give us a different perspective on the mind, helping us to identify the building blocks of mathematical thinking and form interesting connections between mathematics and other expressions of the human mind. To illustrate this point, I now turn to a poem by T. S. Elliot [2]. By analyzing the poem, I hope to show that the same foundational processes of the mind that we have learned about before underlie *poetic imagination,* too. In the next phase, we will show how

---

[1] See "The Logic of Relatives"
   http://www.commens.org/sites/default/files/biblio_attachments/the_logic_of_relatives.pdf

imagination in mathematics can extend our universe of possibilities and return to the point where we opened the book, albeit with a creative twist.

5.2 "What Men Choose to Forget"

Elliot's poem titled *The Dry Salvages* opens with the poet's statement that he doesn't know much about Gods. But he says:

I think that the river.
Is a strong brown god—sullen, untamed and intractable,
Patient to some degree, at first recognized as a frontier;

Let us interpret the poem's opening to see if previously discussed ideas pop up during the analysis. The interpretation will follow an old Jewish tradition, distinguishing four steps/approaches[2] to interpretation:

Peshat (פְּשָׁט): "surface" ("straight") or the literal (direct) meaning.

Remez (רֶמֶז): "hints" or the deep (allegoric, i.e., hidden or symbolic) meaning beyond just the literal sense.

Derash (דְּרַשׁ): from Hebrew *darash*, meaning "inquire" ("seek")—the comparative (midrashic) meaning, as given through similar occurrences.

Sod (סוֹד): "secret" ("mystery") or the esoteric/mystical meaning, as given through inspiration or revelation.

The first step/approach is simple and involves a literal approach to interpreting the text.[3] It is a direct analysis of the text. If a river is mentioned in the opening line of the poem, then it is a river in the most basic sense of the term: "A large natural stream of water flowing in a channel to the sea, a lake, or another, usually larger, stream of the same kind" (Oxford English Dictionary). A river is just a river with all the simple associations that emerge when we think about a river. The same holds true for numbers in mathematics. We may conceive of a number as it is understood. For example, the number "5" means the quantity five. It's a basic, surface-level understanding of the number, where "5" is simply the numeral that follows "4" and precedes "6" in the sequence of natural numbers. Here, "simply" does not denote any absolute quality but is used as a relative term. When we learn to count and manipulate numbers as young children, we are not consciously bothered by anything other than the simple meaning of the number. The number is conceived as that which is countable and no more.

The second approach to interpretation (i.e., **Remez**) proposed by the old Jewish sages is the one in which we *realize* the symbolic meaning of the text. It is the realization, to paraphrase Freud, in which we understand that a cigar is NOT just a cigar.

---

[2] https://en.wikipedia.org/wiki/Pardes_(exegesis)

[3] https://en.wikipedia.org/wiki/Peshat

Here, the river is not just a natural stream of water but a sign of something else. In the first line, Elliot tells us that the river is a God and a frontier. That is, he directs us to inquire about the meaning of "river" as God and frontier.

The **R**emez should remind you of our discussion of numbers. A number, as we learned from Russell and others, is a sign; it is a sign functioning as a name. Russell didn't use this semiotic language. He used the idea of a cardinal number, the number of elements in a set. A number is a class of all sets that are equivalent in the sense of having the same cardinality. For instance, the number "2" is defined as the set of all pairs of objects. This means any set with precisely two elements is an instance of the number 2. From a semiotic perspective, this formalization of a number is a specific instance of naming. The number is, therefore, a sign signifying all sets of the same cardinality, regardless of the particular properties of the elements of the sets.

Recognizing the river as a sign, we inevitably ask what it means. Is there a symmetry between a God and a river? Are we reading a poem by an idol worshipper who believes the river is a God? For the interpretation phase, the Derash, we must assume this is not the case. The poet creates a deep similarity up to identity, in which the river and God are equated. The Derash asks us to understand the meaning of river as God and frontier, using a *structured* method that can help us to bridge the gap in our understanding. This gap lies between the literal and hinted semiotic senses of the signs, which must be bridged to understand the meaning of the similarization. Here, we move to the territory of reasoning, where we try to understand the similarity. The process is equivalent to solving an equation, so the interpretations resulting from this process may be considered possibilities for extending our horizons.

Let us adopt this approach and play around with it. First, we may separate the monadic and dyadic relations in the opening of Elliot's poem. The monadic relation is expressed in the properties used to describe the river God in the poem's opening lines: brown, untamed, etc. The dyadic relation is the identity relation formed by IS_A: the river IS_A God and a frontier. The river is described with the following properties: brown, sullen, untamed, intractable, and patient. Recall how a vector can represent the meaning of a word through the idea of word embedding. I used the list of properties as *monadic relations* to define the word embedding of the river and asked ChatGPT to guess the entity described by these attributes. I find its guess fascinating (my emphasis):

> The vector of words you provided describes a natural entity, specifically an *animal*, with positive and negative attributes. The words 'sullen,' 'untamed,' 'intractable,' 'patient,' 'brown,' 'forgotten,' 'useful,' 'untrustworthy,' 'problem,' 'implacable,' 'destroyer,' 'waiting,' and 'watching' paint a picture of an *animal* that is *wild*, potentially *dangerous*, yet useful and often overlooked.

This guess is remarkable. The river is described through a vector of words whose meaning is that of a *wild animal*. By representing the monadic relations in the text and using them to define the vector of a word embedding for a word that is not known in advance (i.e., a *variable*), we were able to guess the meaning of the river using the most sophisticated LLM technology that currently exists. The structured reasoning approach used here allows us to understand the meaning of the sign "river." Is there an objection to using this technology for interpretation? I see no reason why modern Derash should not use modern tools, both theoretical and technological, that were unavailable to the sages who lived long ago. In sum, we have gained an exciting interpretation by representing the river through a relational approach as a vector in a high-dimensional space and asking ChatGPT to solve the equation and identify the variable.

From the first literal sense of the river (obtained at the Peshat stage), we moved to the river as a sign of a God/frontier (obtained at the Remez stage), and from this sign, we used the monadic relations to interpret the river/God as a wild beast.

The interpretation so far may help us to understand the triadic relation involved in describing the river as a "reminder of what men choose to forget." As readers of the poem, we have no clear idea of what men choose to forget.

We can express the triadic relation as "IS_A (x, y, z)": The river (x) is a reminder (y) of what men choose to forget (z). So, the river is not only a deity but a sign of a thing men choose to forget. What is it that men choose to forget? We read further along that the river "was present in the nursery bedroom" and "In the rank ailanthus of the April dooryard." The rank ailanthus of the April dooryard describes the vigorous and possibly heavenly tree (ailanthus) growing in the garden near the door of a house during April, when spring growth is prominent. The river, described as a *rhythm*, is also present in the following context:

In the smell of grapes on the autumn table,
And the evening circle in the winter gaslight.

I returned to ChatGPT to use its superior word embedding technology and asked it to play the following game with me:

Let us play a guessing game. I have something PRESENT in the following contexts:

1. Nursery bedroom
2. Rank ailanthus
3. Smell of grapes
4. Evening circle in the winter gaslight

Guess what this thing is and the common denominator of the four contexts.

The answer is remarkable (my emphasis):

The common denominator among these contexts seems to be an *evocative presence*— something that triggers a strong sensory or emotional response. One possible guess is *memory*. Memory is something that can be present in all these contexts.

The term "presence" is not easy to grasp. To understand this idea, read "The Mystery of Presence" [3]. The book was not written by a blurred-eyed esoteric hippy loaded with LSD. The author was a talented blind mathematician deeply immersed in the real world and its challenges. Through the book, you may better understand the experience of evocative presence from the ultimate rationalist, who was also a mystic.

Going back to the poem, the divine is described as a memory characterized by an "evocative presence." The evocative presence of the divine reminds us of the presence we choose to forget. A wild, untamed presence in constant flux and in opposition with all we know as civilized human beings. It is a powerful and primordial being, masked and repressed by the constraints of our civilization. This is not an outside force but a presence (i.e., a river), which is described by Elliot as existing "within us," meaning that the divine as an "evocative presence" is within us and "hints of earlier and other creation" (i.e., a memory).

What have we learned so far? We have learned that the same processes discussed as fundamental processes of the mind, those underlying mathematical thinking, can be applied to the interpretation of a poem, as long as we (1) go beyond the literal given meaning, (2) understand the semiotic nature of the signs involved, and (3) apply a structured approach for interpretation and reasoning, which is not exclusively based on our intuition. A relational and processual approach has been applied, too. Using the algebra of relations, I identified the meaning of the word/variable described by the river God's vector (i.e., the word embedding). Surprisingly, it was not the word "river" nor the word "God." However, this is unsurprising when we understand that the signs "river" and "God" are not used literally. The reasoning from the monadic relations to the variable served to bring out the nonliteral meaning of these signs, showing how these signs are similarized.

Using this understanding and applying the same methodology, I tried to understand the deeper meaning of the river as expressed by a triadic relation. Another symbolic meaning was brought to light: the river is a memory described as an evocative presence, a divine primordial force present in every one of us. Here, we already touch upon the final step in the traditional approach to interpretation, known as Sod.

Sod seeks the "esoteric/mystical meaning, as given through inspiration or revelation." The Jewish sages warned against the Sod. It is accessible only to the few, and then only after a long process of learning and contemplation. When interpreting the text, I experienced no revelation and therefore have nothing to say that could be considered within the realm of Sod. However, we may still somehow understand the esoteric experience of Elliot as reflected in and communicated by his poem.

Elliot's revelation is not a revelation of Sod. The entrance to the "temple of knowledge" (mentioned in the preface) was closed long ago. The revelation expressed in the poem is not the one involved in "the disclosure or communication of knowledge, instructions, etc., by divine or supernatural means" (from 1384) but "the disclosure or exposure by a person of something previously unknown."

At their best, poetry and mathematics may share this form of revelation. Through imagination, a realm of unknown possibilities is exposed in a way that evokes a sense of awe. In the following sections, I return to mathematics and show how, through imagination, the realm of complex numbers was revealed, and how a structured form of imagining may extend our understanding of the primary distinction as the root of cognition.

5.3   Negation and the Repression of Zero

Most people I know are not familiar with a concise paper published by Freud [4] and titled *Negation*. Freud considered negation to be the expression of a defense mechanism. In his language: "Negation is a way of taking account of what is repressed." For example, if someone said, "I don't think I'm a homosexual," Freud might have considered this utterance as possibly expressing unconscious homosexual desires, rejected by the conscious mind. Negation is essential as it allows unconscious material to become conscious while still being denied or disputed. However, Freud also says another interesting thing about negation. He suggests that, while affirmation symbolizes the operation of "union" and Eros, negation belongs to the instinct of destruction. Saying "no" separates things. When an infant first refuses, it is a point where she first expresses her individuality as a separate and autonomous human being. The relation between Boole and Freud becomes apparent. Both recognized two opposing forces in action. One force (operation in Boole's terms) is a uniting force that brings things together; the other is a breaking force that separates the whole into parts.

Moreover, when discussing the root of the mind, we learned that a distinction appears when a boundary is formed between inside and outside, between 0 and 1. As 0 symbolizes the realm beyond our grasp, it can only be symbolized by the negation of 1, whence NOT $(1) = 0$. One may find a clear association between my interpretation of Elliot's poem, Freud's negation, and the primary distinction. Symbolizing 0 through negating what there is (i.e., 1) rather than positively affirming its existence is a sign of repression. *The void is beyond our understanding, and it is an evoked presence inducing deep anxiety.* Therefore, we can signify it only through negation. Therefore, 1 and 0 are mutually defined through the most basic negation operation. Moreover, if $x^2 = x$, then the identity of x is defined through 1 and 0, which are the only solutions to this equation. One and zero are not only the values of the primary distinction but also those underlying the most basic meaning of identity. This point will be explained later. But first, I would like to explain how playing with the imagination may open up a new horizon of possibilities. Here, we turn to the imaginary numbers.

5.4   The Imaginary Number as a Kushiya

Imaginary numbers popped up during the history of mathematics as a necessary tool for solving certain equations. Here is the general form of this creature:

$$i^2 = -1 \rightarrow i = \sqrt{-1}$$

where the imaginary unit i is the solution to $i^2 = -1$. Now, the square root of a number is a number that, when raised to the power of 2, gives back the original number. For example,

$$\sqrt{9} = 3$$

because $3 \bullet 3 = 9$. However, there is NO solution to the square root of minus one, as $1 \bullet 1 = 1$, and $-1 \bullet -1 = 1$. Here, we encounter a Kushiya. On the one hand, the imaginary creature is necessary to solve certain problems, but on the other, such a creature seems impossible.

The resolution of the Kushiya was found by a mathematician called Rafael Bombelli.[4] Bombelli took an amazingly creative step, greatly extending the possibilities available to mathematical imagination. The move was simple: the fact that we are unfamiliar with such a creature doesn't mean that such creatures cannot exist. It may just be a different kind of creature that we have not noticed so far, like the platypus. This was a moment of revelation, whether or not Bombelli was aware of it.

The term "imaginary" should not be used exclusively for the creature described above. When the number 0 was first introduced, it was a number that did not reside within the horizon of the first mathematicians. At the time, it was no less imaginary than the imaginary unit i. However, imaginary numbers are different. The imaginary unit cannot be easily imagined with respect to our sensorimotor experience, and the same holds for the other imaginary numbers. Historically, the *use* of imaginary numbers preceded their *understanding*. This fascinating historical fact points to the possible precedence of *use* over *understanding*. Today, excellent explanations[5] are available to establish our understanding of this imaginary creature. First, we should visualize an imaginary dimension perpendicular to the dimension occupied by the real numbers. These two dimensions then span the *complex plane* C. Figure 5.1 shows how it looks, and as an example, Fig. 5.2 shows how the complex number $5 + i$ is represented on the complex plane.

To better understand complex numbers, let us look at the nontrivial way they are multiplied. Multiplication by i corresponds to a 90° rotation of the real number line. However, the product of two *complex* numbers has a deeper meaning. To multiply complex numbers, we use a general procedure for rewriting $(a + b) (c + d)$. The procedure is as follows:

Multiply first terms: ac
Multiple outer terms: ad
Multiply inner terms: bc
Multiple last terms: bd
Add them all up: ac + ad+bc + bd

---

[4] https://www.youtube.com/watch?v=N9QOLrfcKNc
[5] https://www.youtube.com/watch?v=65wYmy8Pf-Y&t=123s

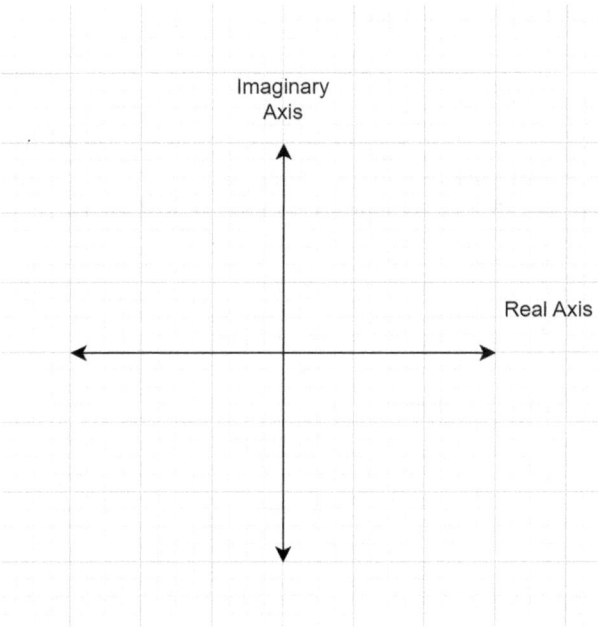

Fig. 5.1   The complex plane. (Source: Author)

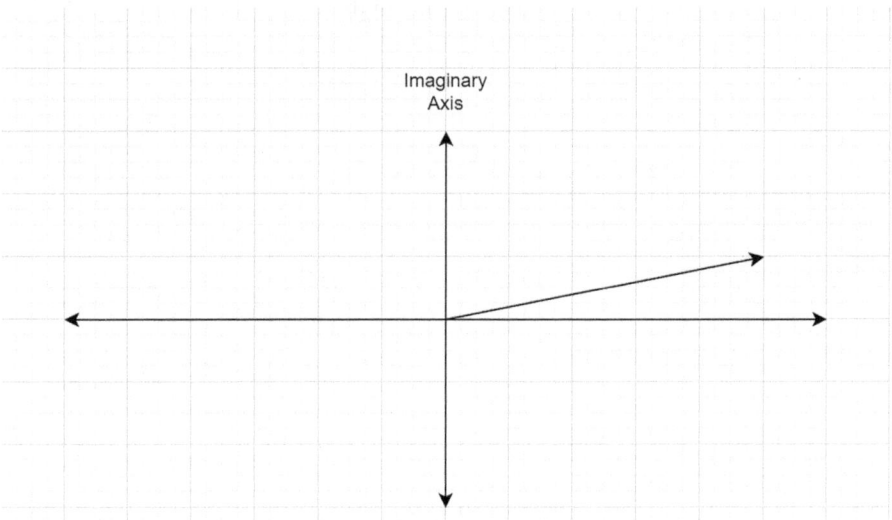

Fig. 5.2   The complex number 5 + i as it is represented on the complex plane. (Source: Author)

For complex numbers, what we get is this:

$$(a+bi)(c+di) = (ac - bd) + i(ad + bc)$$

For instance, (6-5*i*) • (3 + 7*i*) results in 53 + 27*i*. Now, let us examine a simple example to understand the meaning of multiplication. Consider the two complex numbers (3 + *i*) and (3 + 2*i*). Figure 5.3 shows how they are represented on the complex plane.

To reexpress:

$$(3+i)\cdot(3+2i)$$

we multiply the terms as follows:

3•3 = 9
3•2i = 6i
i•3 = 3*i*
i•2i = 2*i*²

and since *i*² is by definition equal to −1, we have 2*i*² = 2•(−1) = −2. Adding the values, we get:

$$9 + 6i + 3i + 2i^2 = 9 + 9i - 2 = 7 + 9i$$

Multiplying the two complex numbers results in the vector shown in Fig. 5.4.

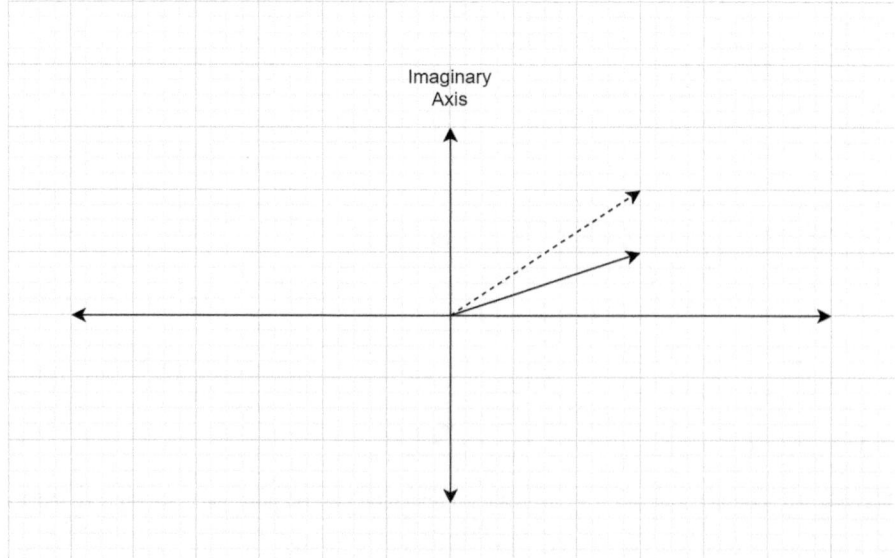

Fig. 5.3 Representing the two complex numbers. (Source: Author)

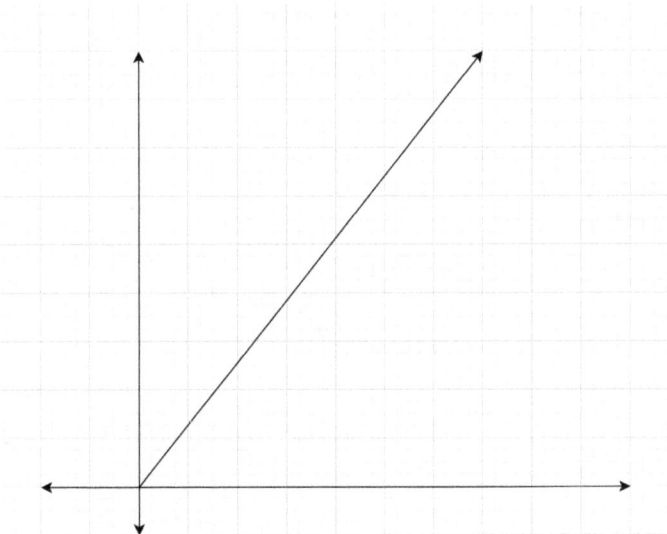

Fig. 5.4 The product vector. (Source: Author)

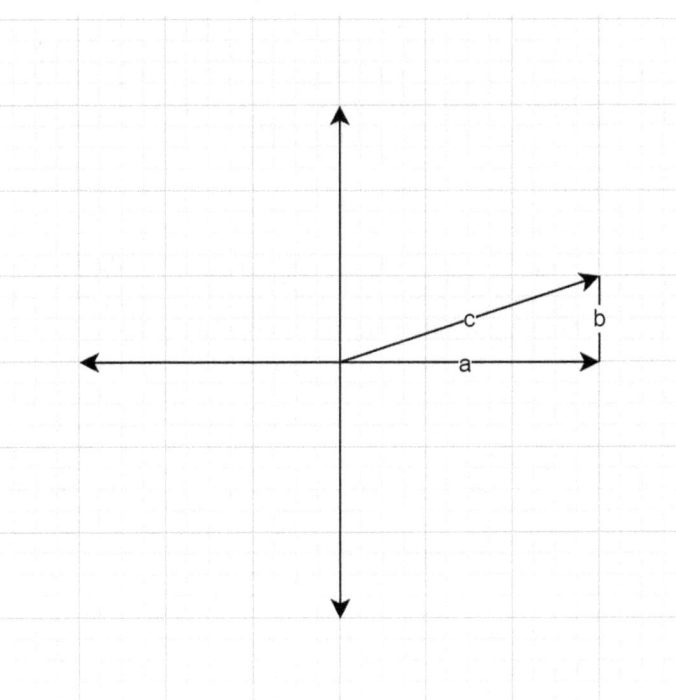

Fig. 5.5 The right triangle associated with the number $(3 + i)$. (Source: Author)

So, what happened when we multiplied the numbers? To understand our results, let us examine the angles made by 3 + i, 3 + 2i, and 7 + 9i with respect to the real axis. We start with 3 + i. Figure 5.5 shows the associated right triangle.

Here,

a is the adjacent
b is the opposite
c is the hypotenuse

Now, the tangent of the triangle is defined as:

$$\tan(\theta) = \frac{Opposite}{Adjacent} = \frac{b}{a}$$

The real part of 3 + i is 3, and the imaginary part is i. Therefore, the angle or the argument of this imaginary number is:

$$\tan^{-1}\left(\frac{1}{3}\right) \approx 18.43°$$

What is the argument/angle of 3 + 2i? By the same reasoning, we find 33.69°, and the angle of the new vector formed by multiplying the two complex numbers is 51.34°. Do you see something interesting? The angle of the new vector formed through multiplication is the *sum* of the angles of the two original vectors.

What about the *length* of the new vector? Is it also the sum of the lengths of the original vectors? Using the Pythagorean theorem:

$$r^2 = x^2 + y^2$$

where r is the length of the vector, x is the real part, and y is the imaginary part. In the case of the first number, we have x = 3 and y = 1. The length or magnitude of the first vector is 3.16. For the second vector, it is 3.60, and for the vector resulting from multiplying the complex numbers, we find 11.40. So the length of the vector formed by multiplying the complex numbers is not the sum of the original lengths but their *product*. When multiplying two complex numbers, we get a new vector whose angle with the real axis is the sum of the angles of the vectors and whose length, or distance from the origin, is the product of the lengths of the vectors. The distance from the origin and the angle with the real axis are the two quantities we need to represent a complex number. This understanding leads us to the *polar representation* of complex numbers:

Magnitude/length (called the modulus)—in the case of our first number, 3.16
Angle (called the argument)—in the case of our first number, 18.43°

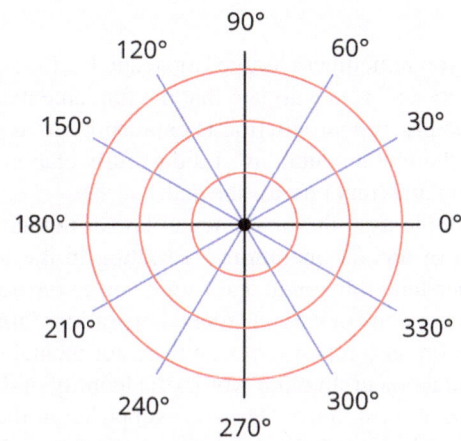

Fig. 5.6 Angles associated with complex numbers in the complex plane. (Source: Wikipedia)

Figure 5.6 shows the angles associated with different complex numbers in the complex plane when they are expressed in polar form.

Let us return to the complex numbers I presented above and their polar form. The first complex number is fully represented by:

$$3.16 < 18.43°$$

Using this representation, we can multiply two complex numbers by multiplying their magnitudes and adding their angles. In the case of our complex numbers, their product is:

$$(3+i)\cdot(3+2i)$$

which can be expressed as:

$$(3.16 < 18.43°)\cdot(3.60 < 33.69°)$$

resulting in the number:

$$(11.40 < 51.34)$$

So far, we have learned how, by imagining imaginary units and complex numbers, we can increase the domain of numbers to include a new territory. However, this is not a mathematics book; the only justification for the discussion so far is to show how complex numbers can be used to imagine possibilities in other domains. This is the aim of the final section, which concludes the book.

5.5   Imagination, Mathematics, and Mysticism

Do you remember George Boole and his fundamental law of thought, where $x \cdot x = x$ and $x^2 = x$? Boole argued that the fundamental law of thought is crucial because the principle of contradiction is entailed by it, as previously explained. When presenting the fundamental law, I didn't fully elaborate on it because I wanted to save the sweet and final bite for the end.

Boole used the fundamental law of thought "for expressing a particular succession of mental operations," resulting in the thing "in itself" [5 , p. 30]. We must understand this point, and I urge you to reread the above-cited text. The law uses a "succession" of certain "mental operations" that result in the thing "in itself." A succession, or a repetition, stabilizes our mental object with no reference to any other mental object. In other words, the identity of the "thing" results from its succession. As Boole explains: "The case supposed in the demonstration of the equation [i.e., $x^2 = x$] is that of *absolute* identity of meaning" (ibid.).

The operation of multiplication expressed by the symbol ($\cdot$) is suitable for this mission because $x \cdot x = x$, but with respect to two values of the variable x, namely $x = 1$ and $x = 0$. This idea is crucial to understanding the primary distinction. In the universe of the primary distinction, identity and stability are guaranteed only if x is 1 or 0. This point is expressed as follows [ibid., p. 36]:

> Thus, it is a consequence of the fact that the fundamental equation of thought is of the second degree, that we perform the operation of analysis and classification, by division into pairs of opposites, or, as it is technically said, by *dichotomy*.

Here, another Kushiya pops up: what would have happened if, instead of using x^2, we had used x^3? Or x^4? What if the succession of mental operations exceeds a single step? Boole answers this question for the case of two steps in the succession of operations, saying [ibid.]:

> Now if the equation in question had been of the third degree, still admitting of interpretation as such, the mental division must have been threefold in character, and we must have proceeded by a species of *trichotomy*, the real nature of which it is impossible for us, with our existing faculties, adequately to conceive, but the laws of which we might still investigate as an object of intellectual speculation.

This is an interesting point. Boole inevitably recognizes the fact that a solution exists for x^3 and that this solution must have three possible values. So what? Boole rejects this possibility on *psychological* rather than *mathematical* grounds by writing, "the real nature of which it is impossible for us, with our existing faculties, adequately to conceive." Why is it difficult to imagine the fundamental law of thought outside the binary world actualized by x^2? Let me explain this point by returning to complex numbers.

What happens when we raise x to the power of 3? We have three possible solutions in this case: $x = 0$, $x = 1$, and $x = -1$. Why three solutions? *The fundamental theorem of algebra* tells us that a polynomial must have as many roots as its highest power. Therefore, when raising x to the power of 3, we have three solutions: 1, 0, and -1. The two binary solutions are those we already found for x^2. They seem comprehensible when we ground them in the world of the first distinction. Therefore, x^2 is a succession that guarantees the identity of x with respect to the two values of the primary distinction: 1 and 0. But how can we understand the value of "-1"? It is not a part of the binary logic formed by the primary distinction, and it doesn't correspond to anything in Boole's world of logic. We understand that a proposition may be true or false, 1 or 0, but what could it mean to say that it is "-1"?

Boole's fundamental law of thought assures that there are two basic values that, through a single mental operation (i.e., a single succession), can guarantee the identity of a thing (i.e., x). It is a world grounded in duality, the world of the primary distinction, of opposition theory, and Gnostic dualism. Moreover, the two values, 1 and 0, are values of the most basic "system" of the mind: the primary distinction. Moreover, these two elements form a *closure*, as we can get from 0 to 1 and from 1 to 0 through the operation of negation. This is a fully understandable and mathematically justified system. But is there a single and simple arithmetic solution for transforming each *triad* member (i.e., 1, 0, and -1) to each of the others? Can a single operation help us to assure this triad's closure? The answer is negative, so we can understand some of Boole's concerns.

However, we may be able to better understand the meaning of identity with respect to x^n using the idea of complex numbers. When $x^3 = x$, we solve $x^3 - x = 0$ in the complex number plane, obtaining the following solutions:

$x = 0$, and the solution in polar form is $0 < 0°$

$x = 1$, and the solution in polar form is $1 < 0°$

$x = -1$, and the solution in polar form is 1, π radians (since -1 lies on the negative real axis). When we speak of an angle of π radians from the positive real axis, we refer to an angle that represents a half-turn (or 180 degrees) counterclockwise from the positive direction of the real axis.

In this case, a single and simple transformation equivalent to the NOT gate does not exist for this triadic set of values. The Kushiya remains unsolved. So, let us use our imagination to move forward.

Boole's binary world seems shaken by the introduction of more solutions, which leads us astray from our basic intuitions and the binary root of the mind. So, let's ask a *different question*. Assuming a world formed by the primary distinction and its binary values, what are the solutions to the equations $x^3 = 1$ and $x^3 = 0$? In other words, instead of asking which values remain the same given the operation, we ask which values can undergo a succession of transformations that result in the two values composing the primary distinction. For $x^3 = 0$, there is only one solution, namely "0." Nothing leads us only to nothing. Nothing is a mystery, and the way something emerges out of nothing is both a fact and a mystery. For the equation $x^3 = 1$, things become more interesting. What is the x that, when raised to the power of 3, results in 1? What number, through three successions, will bring us to 1?

Let us solve the equation $x^3 = 1$. The first solution is trivial: $x = 1$. What about the other solutions? We can rewrite the equation in the form $x^3 - 1 = 0$ and solve it using complex numbers. Regarding the polar form, 1 can be represented on the complex plane as having a real value of 1 and an angle of 0° or 360°. Now, if the product x * x * x is 1, x must have "1" as its magnitude. This is because we have learned that the magnitude of the new vector is the product of the original magnitudes, and since only 1 * 1 *1 = 1, the magnitude of our *solutions* must be "1." What about the angles of our solutions? The angle of the product is the sum of the original angles. Since the angle of the product is 360°, the second solution can have an angle:

$$\frac{360°}{3} = 120°$$

Hence, the second solution has a polar form of 1 < 120°, which can be represented as shown in Fig. 5.7.

We see that the second and nontrivial solution involves a number represented in complex form. This solution can be understood as a *rotation* performed on a number (i.e., a vector) existing on the complex plane. Here is how it works:

$$\left(1 < 120°\right)\left(1 < 120°\right)\left(1 < 120°\right) = \left(1 < 240°\right)\left(1 < 120°\right) = \left(1 < 360°\right)$$

This means that when we multiply this number by itself three times, we return to the point where we started (i.e., 1). There is also a third solution, 1<−120°, which works like the second solution but with a clockwise rotation.

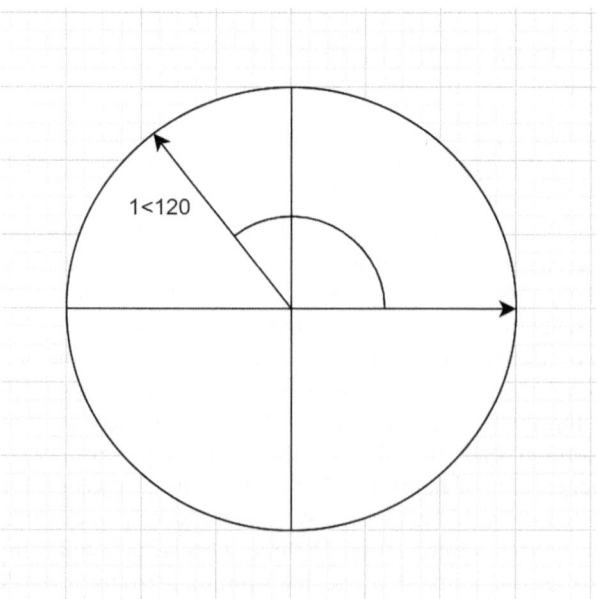

Fig. 5.7 The second solution expressed as a rotation. (Source: Author)

As we learned in Chap. 1, Boole's law of thought involves *time*, *repetition*, and *transformation*. Here, we see these ideas expressed with respect to the solution of the equation $x^3 = 1$. In dynamic terms, the above solutions can be described as transformations performed on a vector in the complex plane. Why is this interesting? For Boole, the meaning of the most basic repetition is two values that are mutually defined by negation: $1 = $ NOT (0) and $0 = $ NOT (1). For higher-order forms, the solutions are not intuitive. We don't understand how the identity of the two primary values can be assured for longer successions of the operation. However, by revising the question to deal with the identity of 1 and representing the solutions using the complex plane, we can now understand them in terms of transformations defined by rotation. The only variable that changes is the *angle*, which depends on the power of x and can be either anticlockwise or clockwise. In other words, the solution to $x^n = 1$ has now become highly intuitive. It is determined by rotations in the complex plane. From a world governed by a primitive relation (i.e., NOT) and two discrete values (1 and 0), we have moved into a world where the identity of what exists (i.e., 1) is guaranteed by easily conceivable transformations in the complex plane. Boole, bothered by the nonintuitive expressions of his law of thought in higher powers of x, can now rest in peace.

The root of the mind lies in the primary distinction. This distinction is described through the values of 1 and 0. While nothing exists in and of itself, the identity of a thing (i.e., 1) can be imagined as long as we keep juggling the balls, metaphorically speaking. We can imagine the succession of x's to be infinitely long. From the perspective of finite minds, things exist as long as they form a succession in our minds. From the perspective of the infinite, things can exist forever.

Is there something outside our minds and their represented realities? Here, we touch on the point where the mind, mathematics, and mysticism form a nexus. When we ask why something exists rather than nothing, we might answer as follows: something exists as long as it is imagined in time through repetition, transformation, and motion. The rest is silence.

Points to Note
- **The algebra of relations**: is crucial for understanding the mind.
- **Poetry and mathematics**: share the feeling of awe we get when realizing possibilities.
- **Imagination and complex numbers**: complex numbers show how imagination extends our minds.
- **Structured imagination**: paradoxically, imagination is productive as long as it is constrained.
- **The complex plane and identity**: The complex plane allows us to model intuitively the way a thing's stability is maintained for x^n, when $n > 2$.
- **Existence through imagination**: existence is assured as long as it is imagined in time through repetition, transformation, and motion.

References

1. Tesnière, L.: Elements of Structural Syntax. John Benjamins Publishing Company, Amsterdam (2015)
2. Elliot, T.S.: Four Quartets. Faber & Faber, London (1944)
3. Amit, I.: The Mystery of Presence. Aliat Gag, Tel Aviv (2013)
4. Freud, S.: Negation. Int. Rev. Psychoanal. **6**, 367–371 (1925)
5. Boole, G.: An Investigation of the Laws of Thought. Dover Publications, New York (1854)

Chapter 6
Epilogue

Abstract The epilogue reflects on concluding the book, comparing it to summarizing a concert. I summarize the book as a journey through mind, mathematics, and imagination, highlighting their profound interconnectedness. Ultimately, the book aims to deepen our understanding of the mind and mathematics, inspiring further exploration of the intricate dance of distinction, signification, and transformation within the human cognitive landscape.

I find it strange to write a concluding chapter for this book. Imagine a concert where the orchestra and the choir play the St. Matthew Passion by Bach. When the concert ends and the audience applauds, someone gets on his feet, shouting in protest: "But what about the summarizing chapter! Please summarize what we have listened to in this concert!" Well, if you haven't understood it so far … However, for the reader seeking "closure," let me provide a short epilogue instead of a concluding chapter.

This short book aims to take the reader on a journey into the mind, mathematics, and imagination. During the journey, I hope to have exposed the profound interconnectedness that defines our cognitive landscape and its nontrivial aspects. At the heart of this landscape lies the primary distinction, a fundamental concept for understanding the mind and mathematics. This concept is more than a binary divide; it is the very seed of the mind, the first step toward forming identity and understanding similarities and differences.

As we traveled through the different expressions of the mind, signs emerged as pivotal elements that empower the mind by supporting abstraction and imagination, acting as values of functions and attentive acts. There is no mathematics without signs. We achieve mathematics, like all higher cognitive functions, through signs. Therefore, signs are not simple technical devices for communication. This is not a trivial issue. The connection between "language" and "thought" suffers from a bad reputation, given the shameful errors of misunderstanding and deliberately biased opinions propagated by overzealous students of the already zealous Chomsky.[1] There is no mathematics without signs and for a good reason. Through repetition, a

---

[1] For a discussion of language and thought, read Neuman, Y.: *Introduction to Computational Cultural Psychology*. Cambridge University Press, Cambridge (2014).

Y. Neuman, *Mindmatics*, Mathematics in Mind,
https://doi.org/10.1007/978-3-031-74955-1_6

sign in itself, these signs support the basic concept of identity, where some stability is gained over the underlying dynamics of things.

Boole's fundamental law of thought encapsulates this idea, highlighting the interplay between repetition, similarity, and identity. Moreover, during my exploration of repetition, we encountered the concepts of fixed points and invariant sets. These islands of stability express the way the mind stabilizes an underlying flux of events. Recall Heraclitus's famous aphorism: "No man ever steps in the same river twice, for it's not the same river and he's not the same man." Mind and mathematics show us how to step into the same river more than twice by stabilizing the flux. In this context, we understand that repetition, as expressed in Boole's fundamental law of thought, is not merely a mechanical reproduction process but a structural principle expressing similarity, symmetry, and transformation. This understanding led me to adopt an event-centered approach, in which I emphasized the importance of studying the mind through events, interactions, and transformations, recognizing that our mental and mathematical endeavors are deeply intertwined.

With repeated reference to art, I also showed how some aesthetic experiences arise from repetition with variation. I also showed how different artistic expressions of the mind, from children's stories to poetry, expose the logic of the unconscious. I argued that the unconscious, with its symmetrical dynamic, plays a crucial role in shaping our imagination, just as it underpins the logical frameworks of mathematics. Discussing the Principle of Symmetry governing the Unc, I pointed out the connection between repetition and dimensionality reduction, revealing how collapsing different signs into lower-dimensional spaces can create new forms of identity and understanding. Opening up the realm of possibilities through the unconscious led me to discuss imaginary and complex numbers. Through the lens of complex numbers, I showed how structured imagination deepens our understanding of the primary distinction, highlighting the elegance of rotation on the complex plane as a metaphor for understanding the stabilization of existence.

The idea that something exists as long as it is imagined in time, through repetition, transformation, and motion, encapsulates the essence of our journey through mind and mathematics. In this symphony of the mind, mathematics, and imagination, we uncover the intricate dance of distinction, signification, and transformation. This exploration deepens our understanding of the mind and mathematics, inspiring us to continue probing the mysteries of the human mind, where the realms of thought, abstraction, and imagination converge harmoniously. The rest is silence.

Author Index

A
Alexander, C., 27
Aristotle, 13, 46
Auster, P., 35

B
Bateson, G., 47, 48, 64
Bohm, D., 5, 7, 14, 25, 48
Boole, G., 3–11, 13, 14, 23–26, 33, 34, 38, 45,
    81, 88, 89, 91, 94

C
Chouchani, M., 3, 35, 48

E
Elliot, T. S., 75–78, 80, 81

F
Freud, S., 5, 34–36, 59–64, 77, 81

H
Halmos, P., 41, 43, 44
Harnad, S., 42

Hewitt, P. G., 25

J
Jennings, H. S., 17

K
Kauffman, L., 4, 23
King, S., 36

L
Leibowitz, Y., 5, 6

M
Matte-Blanco, I., 60–62
McCarthy, C., 31, 50

P
Peirce, C. S., 76
Poincaré, H., 21

R
Russell, B., 15, 16, 62, 78

S
Spencer-Brown, G., 2, 3, 8–10, 16, 23, 25

T
Tesnière, L., 48, 76
Thomas, D., 23, 24, 36, 38, 70, 72

V
von Foerster, H., 3, 34

W
Whitehead, A N., 15, 22, 23, 31, 32, 45,
    47, 48

Subject Index

A
Abstractions, 4, 15, 16, 18, 21, 23, 48, 93, 94
Aggregate, 4, 7, 11, 13, 47
Alice's adventures, 62
Analogies, 9, 15, 16, 18
Association-by-sense, 15
Asymmetry, 55, 60, 61
Attentive acts, 5, 18, 93

B
Belonging, 42, 43
Bi-logic, 62, 65
Boundaries, 2, 3, 9–13, 27, 42, 81
Brouwer's fixed point theorem, 26

C
Categories, 4, 6, 10, 15, 28, 42, 45, 62
Categorization, 42
Category theory, 4
Center, 26, 27, 38
ChatGPT, 65, 78, 79
Classes, 4–11, 13–15, 17, 18, 33, 46, 48, 49,
        62, 65, 66
Complex numbers, 2, 80, 82–84, 86–91, 94
Condensation, 65, 66, 68
Co-product, 6, 7
Counting, 15, 24

D
Danesi, M., 19
Derash, 77–79

Dictionaries, 27–28, 55, 66, 77
Differences, 1, 13–18, 21, 23, 27, 29–31, 36,
        42, 43, 45, 47–49, 60, 65–68, 71,
        72, 75, 93
Dimensionality reduction, 68–73, 94
Distinctions, 1–6, 8–18, 23, 25, 26, 28, 29,
        32–34, 37, 38, 41, 48, 49, 80, 81,
        88, 89, 91, 93, 94

E
Embodiment, 43–44
Empty set, 4, 9, 11, 47
Ensembles, 6–7, 16
Equality, 6, 26, 29, 30, 32–34, 41, 45, 47
Etymology, 29, 37, 63
Event-centered approach, 52, 55–57, 94
Evocative presence, 79, 80

F
Fixed points, 26, 27, 34, 35, 38, 94
Functor, 15
Fundamental law of thought, 7–9, 18,
        88, 89, 94
Fundamental theorem of algebra, 88

G
Genesis, 2, 12

H
Hermeneutic circularity, 27

I

Identities, 1, 6–9, 11, 13, 18, 23, 25, 26,
         29–30, 32–35, 37, 38, 41, 49, 61,
         65, 66, 68–70, 78, 81, 88, 89,
         91, 93, 94
Imaginary numbers, 5, 31, 81–87
Imagination, 15, 18, 31–32, 47, 54, 56, 57,
         59–72, 75–91, 93, 94
Inference-through-substitution, 37, 38
Information compression, 25
Information entropy, 31
Initial object, 11
Intersection, 7, 9, 76
Invariant sets, 26–28, 34, 38, 94
Iterated function, 25–27, 33, 38

K

Kernel, 27–28, 34
Kushiya, 3, 4, 11, 35–38, 45, 48, 61, 64, 81–89

L

Lakoff, G., 57
Lattice, 10, 11, 13
Linear algebra, 2, 49, 54, 67, 68

M

Mapping, 4–6, 10, 11, 15, 16, 26, 27, 33
Matrices, 48–57, 67, 68
Memories, 17, 23, 34, 35
Metaphor, 36, 43, 94
Modus ponens, 23
Morphism, 4, 15, 76
Mysticism, 10, 75–91

N

Negation, 10, 11, 23, 81, 89, 91
Neuman, Y., 19, 39, 57
Nonself, 1, 5, 9
Nothing, 2, 3, 6, 8–12, 17, 22, 23, 25, 33, 34,
         42, 46, 53, 55, 61, 63, 80, 89, 91
Numbers, 7–9, 15, 16, 18, 21, 24, 30, 32–34,
         41, 46, 49–53, 56, 61, 67, 76–78,
         82, 85–87, 89, 90

O

Observer, 3, 5, 12, 34
Oppositions, 1, 3, 9, 11, 18, 37, 38, 80, 89
Order, 23, 35, 43–45, 52, 55, 62, 66

P

Peshat, 77, 79
Poetry, 23, 24, 33, 36–38, 70, 72, 94
Post Traumatic Stress Disorder
         (PTSD), 35
Potentiality/actuality, 46, 47
Principle of contradiction, 9, 88
Principle of generalization, 62
Principle of symmetry, 61, 64, 94

R

Redundant, 25
Relational, 16, 23, 49, 55, 56, 79, 80
Relations, 5, 15–18, 23, 31, 42–44, 55, 56, 64,
         75, 76, 78–81, 91
Remez, 77–79
Repetitions, 8, 18, 23–30, 32–38, 41,
         49, 59, 60, 69–73, 88, 91,
         93, 94
Rhymes, 36, 37
Russell's paradox, 62

S

Self-reference, 4, 48
Semiotics, 6, 18, 36, 78, 80
Set theory, 6–7, 41–44
Signs, 3–7, 10, 11, 13–16, 18, 21–23, 25,
         27–29, 31–34, 36, 45–50, 66,
         70, 93, 94
Similarities, 13–18, 21, 23, 24, 29, 36–38, 41,
         43, 46, 49, 59, 60, 63, 65, 66,
         68–72, 78, 93, 94
Sod, 77, 80
Structures, 1, 10, 13, 15, 22, 23, 27, 30, 37, 38,
         41, 55, 60, 69, 71, 72
Substitutive, 45, 50, 63
Syllogism, 22, 23
Symbolic father, 66
Symmetry, 9, 30, 33, 38, 59–61, 69, 72,
         78, 94
Synecdoche, 63, 64

T

Terminal object, 34
Times, 6, 8, 11, 13, 18, 22, 24, 25, 36, 37, 62,
         64, 66, 69, 70, 94
Transformation matrix, 67, 68
Transformations, 7, 8, 15, 18, 26, 30, 33–35,
         38, 46, 48–57, 61, 65, 67, 68,
         89, 91, 94

U
Unconscious, 9, 13, 36, 38, 59–72, 81, 94
Union, 6, 7
Universe, 2–5, 9–14, 34, 77, 88

V
Values, 2–6, 8–10, 18, 22, 23, 25, 33, 34,
   45–47, 50, 52, 53, 60, 70–72, 93

Vectors, 7, 49–57, 65–71, 86
Vector spaces, 7, 49

W
Word embedding, 66, 78–80